Trafficking of Cardiac Ion Channels

Trafficking of Cardiac Ion Channels—Mechanisms and Alterations Leading to Disease

Editor

Marcel Verges

MDPI • Basel • Beijing • Wuhan • Barcelona • Belgrade • Manchester • Tokyo • Cluj • Tianjin

Editor
Marcel Verges
University of Girona and Girona Biomedical Research Institute (IDIBGI)
Spain

Editorial Office
MDPI
St. Alban-Anlage 66
4052 Basel, Switzerland

This is a reprint of articles from the Special Issue published online in the open access journal *Biomolecules* (ISSN 2218-273X) (available at: https://www.mdpi.com/journal/biomolecules/special_issues/trafficking_ion_channels).

For citation purposes, cite each article independently as indicated on the article page online and as indicated below:

LastName, A.A.; LastName, B.B.; LastName, C.C. Article Title. *Journal Name* **Year**, *Article Number*, Page Range.

ISBN 978-3-03943-472-5 (Hbk)
ISBN 978-3-03943-473-2 (PDF)

Contents

About the Editor

Marcel Verges (Associate Professor). My doctoral thesis (1994), performed at the Univ. of Barcelona, dealt with the endocytic pathway in polarized cells using rat liver as a model. Then, I began a postdoctoral stay at the Univ. of California San Francisco. There, I focused first on lipoprotein uptake and transport in hepatocytes, using rat and knockout mice as model systems. Next, I studied regulation of transcytosis in polarized cells. My results contributed to the understanding of polarized trafficking, demonstrating the role by retromer in mediating polarized protein sorting at the endosome. In 2004, through a Ramón y Cajal contract, I started up my own lab at the Centro de Investigación Príncipe Felipe (CIPF) in Valencia. There, I continued with my efforts to elucidate the mechanism by which retromer regulates polarized sorting. In a supervised doctoral thesis during this period at the CIPF, we showed results that helped to understand the link of sorting defects by retromer and generation of the β-amyloid peptide, as cause and aggravating factor of Alzheimer's disease. We then contributed with additional data on localization and function of retromer in polarized trafficking. In 2010, I obtained a position as tenure-track Lecturer at the University of Girona (UdG) School of Medicine, and I also joined GenCardio, as a senior researcher, at the Girona Biomedical Research Institute (IDIBGI). Through collaborative work, we showed for the first time the association with Brugada syndrome (BrS) of SCN2B, the gene encoding the voltage-gated sodium (NaV) channel β2 subunit; BrS patients display cardiac arrhythmia that can lead to sudden death. Subsequent accomplishments have been reached by supervision of various master and undergraduate students, and specially a PhD student now at the stage of thesis defense. First, we have provided an explanation on why β2 D211G—the BrS-associated mutant—is defective in promoting localization to the cell surface of NaV1.5, the major cardiac NaV isoform. More recently, we have shown the key importance of β2 N-glycosylation in its trafficking to the plasma membrane, and in promoting proper NaV1.5 localization. Finally, we have found that β2 is palmitoylated, which increases its affinity for cholesterol-enriched lipid raft domains, influencing its polarized trafficking. These mechanisms are fundamental for cardiac function, since possible alterations can affect cell excitability and electrical coupling of the heart. In July 2019, I was appointed permanent faculty as Associate Prof. for the University of Girona (UdG) at the Medical Sciences Dep.

Preface to "Trafficking of Cardiac Ion Channels—Mechanisms and Alterations Leading to Disease"

This volume contains the articles published in the Special Issue of Biomolecules entitled "Trafficking of Cardiac Ion Channels—Mechanisms and Alterations Leading to Disease" and includes 7 reviews and an original research article. All contributing authors were invited to submit their manuscripts to this Special Issue, either as a review or a research work. Their submitted manuscripts were peer-reviewed in order to appear in this volume. Authored by some of the leading scientists in the field, this volume provides up-to-date information and directions for future research. The articles address mechanisms regulating trafficking and localization of proteins implicated in cardiac function, with special emphasis on ion channels. They present the effects of genetic variants conferring risks of channelopathies and cardiomyopathies linked to sudden cardiac death. More generally, there is also discussion on cellular dysregulation due to alterations in protein processing and targeting as an attempt to improve our understanding of the etiology of human disease. More specifically, this volume includes articles dealing on the following topics: gap junction proteins in regulation of cell cycle; trafficking of the cardiac voltage-gated sodium channel (NaV1.5) and its associated β subunits; cell-adhesion role of NaV1.5-associated β subunits at the cardiomyocyte intercalated discs to promote ephaptic conduction; structural analysis of potassium channel KV11.1 toward identification of missense gene variants conferring risk of channelopathies; structural mapping of inward rectifier potassium (Kir) channels toward identification of mutation clusters responsible for trafficking defects; role of cytoskeleton-associated ankyrin-B in localization of ion channels and other molecules, along with its dysfunction associated with disease; trafficking of cation channels of the vanilloid subfamily of transient receptor potential (TRP) channels based on their molecular interactions in cells; and a broad review on how alterations in protein folding can cause their mislocalization and aggregation, presenting the potential of pharmacological chaperones to maintain protein quality control, thereby preventing cell dysfunction and disease.

Marcel Verges
Editor

 *biomolecules*

Article

An Alternatively Translated Connexin 43 Isoform, GJA1-11k, Localizes to the Nucleus and Can Inhibit Cell Cycle Progression

Irina Epifantseva [1], Shaohua Xiao [1], Rachel E. Baum [1,4], André G. Kléber [2], TingTing Hong [1,3] and Robin M. Shaw [4,*]

[1] Smidt Heart Institute, Graduate Program in Biomedical Sciences, Cedars-Sinai Medical Center, Los Angeles, CA 90048, USA; iepifantseva@gmail.com (I.E.); sxiao@mednet.ucla.edu (S.X.); rachel.e.baum@utah.edu (R.E.B.); TingTing.Hong@cshs.org (T.H.)

[2] Department of Pathology, Beth Israel & Deaconess Medical Center, Harvard Medical School, Boston, MA 02115, USA; akleber@bidmc.harvard.edu

[3] Department of Medicine, University of California Los Angeles, Los Angeles, CA 90048, USA

[4] Nora Eccles Harrison Cardiovascular Research and Training Institute, University of Utah, Salt Lake City, UT 84112, USA

* Correspondence: Robin.Shaw@hsc.utah.edu; Tel.: +(801)-587-5845

Received: 7 January 2020; Accepted: 15 March 2020; Published: 20 March 2020

Abstract: Connexin 43 (Cx43) is a gap junction protein that assembles at the cell border to form intercellular gap junction (GJ) channels which allow for cell–cell communication by facilitating the rapid transmission of ions and other small molecules between adjacent cells. Non-canonical roles of Cx43, and specifically its C-terminal domain, have been identified in the regulation of Cx43 trafficking, mitochondrial preconditioning, cell proliferation, and tumor formation, yet the mechanisms are still being explored. It was recently identified that up to six truncated isoforms of Cx43 are endogenously produced via alternative translation from internal start codons in addition to full length Cx43, all from the same mRNA produced by the gene *GJA1*. GJA1-11k, the 11kDa alternatively translated isoform of Cx43, does not have a known role in the formation of gap junction channels, and little is known about its function. Here, we report that over expressed GJA1-11k, unlike the other five truncated isoforms, preferentially localizes to the nucleus in HEK293FT cells and suppresses cell growth by limiting cell cycle progression from the G_0/G_1 phase to the S phase. Furthermore, these functions are independent of the channel-forming full-length Cx43 isoform. Understanding the apparently unique role of GJA1-11k and its generation in cell cycle regulation may uncover a new target for affecting cell growth in multiple disease models.

Keywords: connexin43; trafficking; internal translation; cell cycle

---

1. Introduction.

Gap Junction (GJ) channels are formed by connexin proteins, which assemble as hexamers on the cell membrane and dock with the hexamer in an adjacent cell, creating a channel for intercellular communication [1]. Connexin 43 (Cx43) is the most prevalent gap junction protein in the heart and is widely expressed in most mammalian organs and cell types [2]. Dysregulation in Cx43 protein expression is associated with cardiac arrhythmias [3]. Studies have identified numerous non-canonical roles of Cx43 that are not described by gap junction channel activity, including in ischemic injury protection [4,5], cancer [6–9], wound healing [10], muscle differentiation [11], organ morphogenesis [12,13], and cell migration in embryonic development [14]. With specific regard to cancer, Cx43 can function as a tumor suppressor [6,15,16]. A reduction in Cx43 expression is observed in tumor cell lines [17], and reduced Cx43 has been proposed as a biomarker of malignant tissue [8,18–20].

Few studies have been able to reconcile how the gap junction Cx43 channel function can influence these apparently channel-independent roles of Cx43 protein.

Cx43 is encoded by the *GJA1* gene, which has a single coding exon [21]. Therefore, *GJA1* cannot provide protein diversity by means of alternative splicing of exons. However, it has recently been found that several small isoforms of Cx43 are produced by alternative translation, with methionines within the coding exon serving as internal translation start sites [22,23]. The alternatively translated small isoforms lack the N-terminal portions of Cx43 upstream of the initiating start sites, while retaining the remaining downstream C-terminus portion. There are six internal methionines and a total of six truncated protein isoforms can be produced from Cx43 mRNA with expression levels varying between isoforms and across cell types [22]. We have previously reported that the size of the most predominantly expressed isoform is 20kDa, thus termed GJA1-20k. It has been identified that GJA1-20k functions in the forward trafficking of full-length Cx43 to the plasma membrane to form gap junction channels [22,24] and is necessary for actin stabilization [24]. In addition to its role as a chaperone [22,24], GJA1-20k is upregulated in response to hypoxic [25] and oxidative stress for preventing mitochondrial fragmentation [26] and to mediate ischemia preconditioning protection [24,27]. Xenopus derived GJA1-20k has also recently been identified to localize to the cell nucleus, functioning as a transcription activator of N-Cadherin [28].

Nucleus localization of one or more of the smaller isoforms provides the opportunity to affect cell cycle. The overexpression of various portions of the C-terminus fragment of Cx43 (Cx43 CT) has been shown to occasionally localize to the nucleus [29,30] and is implicated in the suppression of cell proliferation [29,31,32]. Now that the six endogenous truncation isoforms have been identified [22], previous studies exploring various lengths of the C-terminus can be placed in the context of proteins that can be endogenously generated by alternative translation.

In this study, we perform an expression analysis of all six mammalian isoforms and make the surprising observation that only mammalian GJA1-11k, and not GJA1-20k or any of the other Cx43 isoforms, is preferentially enriched in the nucleus of mammalian cells. Moreover, we find that GJA1-11k can interfere with cell growth by limiting cell cycle progression in the G_0/G_1 phase.

2. Materials and Methods

2.1. Cell Culture

HEK293FT (Thermo Scientific) cells at low passages were grown on petri dishes at 37 °C in a humidified atmosphere with 5% CO2 in Dulbecco's Modified Eagle' Medium (DMEM, Thermo Scientific) high glucose with 10% fetal bovine serum (FBS), nonessential amino acids, sodium pyruvate (Thermo Scientific), and antibiotics Mycozap-CL (Lonza). For the expression of targets, cells were plated down 24h prior to transient transfection at a density indicated in each experiment. Transient transfection was carried out using either Lipofectamine™ 2000 (Thermo Scientific) or FuGENE®HD (Promega). Transfection by both reagents consistently resulted in 75–80% transfection efficiency, assessed by Flow Cytometry. The cells were used for assay after 48 hours of transfection unless otherwise stated in the figure legends.

2.2. RNAi Interference

Chemically synthesized siRNAs (Thermo Scientific) to knockdown *GJA1* mRNA (sequence 5′ to 3′: GG GAG AUG AGC AGU CUG CCU UUC GU; HSS178257) and Stealth™ RNAi with medium GC content as a negative control (scramble, cat. # 12935-112) were used. HEK293FT cells at the density 2×10^6 were transfected with 100nM of either siRNA or the scramble control according to the manufacturer's reverse transfection protocol using Lipofectamine™ RNAiMAX (Thermo Scientific). The cells were manually counted at 24, 48, 72, 96, and 120 h after transfection.

2.3. Molecular Biology

Human *GJA1* encoding full-length and small isoforms were obtained from Open Biosystems and cloned into pDONR/221 using Gateway BP cloning to generate entry clones (BP clonase II; Thermo Scientific). Destination vectors (pDEST) encoding C-terminal V5-, HA-tagged and C-terminal NES (nuclear export signal) V5-tagged proteins were subsequently made using entry clones and Gateway LR cloning (LR clonase; Thermo Scientific). All constructs for mammalian cell expression are driven by the cytomegalovirus (CMV) promoter and include downstream internal mutated methionine-to-leucine start sites of Cx43 to ensure single isoform expression. Mutagenesis was carried out with QuickChange Lightning Mutagenesis Kit (Agilent) according to the manufacturer's protocol.

2.4. Immunofluorescence Staining

HEK293FT cells were plated on 35-mm fibronectin-coated glass bottom culture dishes (MatTek Corporation, cat #P35G-1.0-14-C) 24 hours before transfection. The cells were transfected with 2 mg of two plasmids—the negative control pcDNA3.2-GW-CAT-V5 and the positive control pcDNA3.2-GFP-V5—to verify the transfection efficiency, wild type pcDNA-GJA1-WT-V5, and all of the alternatively translated isoforms of the Cx43 protein (pcDNA-GJA1-43k-V5, pcDNA-GJA1-32k-V5, pcDNA-GJA1-29k-V5, pcDNA-GJA1-26k-V5, pcDNA-GJA1-20k-V5, pcDNA-GJA1-11k-V5, pc DNA-GJA1-7k-V5) and pcDNA-GJA1-11k-NES-V5. At 36h post transfection, the cells were washed in PBS and fixed in 4% paraformaldehyde for 15 min at RT. The cells were permeabilized in 0.1% Triton in PBS and blocked in 5% goat serum for 1 h and were then subsequently incubated with primary antibodies for 3 h at RT: rabbit polyclonal anti-Cx43, raised against a 17-residue peptide from the C-tail of Cx43 (1:1000, Sigma), and mouse monoclonal anti-V5 (1:50, Sigma). After washing 3 times with 0.1% Tween-20 in PBS, the cells were incubated with secondary antibodies: goat anti-mouse IgG conjugated to Alexa Fluor 488 (1:500, Thermo Scientific) and goat anti-rabbit IgG conjugated to Alexa Fluor 555 (1:500, Thermo Scientific) for 1 h at RT. The nuclei were detected by Hoechst 33342, trihydrochloride, trihydrate (1 µg/mL, Thermo Scientific) staining for 5 min at RT and were then mounted to ProLong Gold Antifade reagent (Thermo Scientific).

2.5. Image Processing

Images were taken using a Nikon Eclipse T*i* microscope with a 100×/0.75 Plan Apo objective and a Yokogowa CSU-X1 spinning-disk confocal unit with 350, 486, 561-laser units, and an ORCA-Flash 4.0 Hamamatsu camera (C11440), controlled by NIS Elements software and analyzed using Adobe Photoshop. The HEK293FT cells were imaged at z-depth increments of 0.3 µm. One plane image was used for quantifications in ImageJ (National Institute of Health, Bethesda, MS, USA) and the focused image (maximum projection intensity) of z-stacks was used for the publication. Fluorescence intensity profiles were generated and quantified by Image J in at least 20 cells per sample. The ratio of nuclear/cytoplasmic fluorescence intensity of Cx43 small isoforms normalized to the region of interest (ROI) detected by both antibodies (anti-V5 and anti-Cx43) was quantified. Each image was background-subtracted using a rolling ball radius of 50 pixels.

2.6. Cell Count and Cell Cycle Assay

The cells were plated down at the density 2×10^6 cells per 100mm petri dish and were exposed to serum starvation in complete media with 0.2% FBS for 48 h to ensure maximum cell cycle synchronization without cell cycle arrest. Then, the cells were transfected in 10% serum-supplemented media with various plasmids of interest using FuGENE®HD (Promega) according to manufacturer's protocol. The concentration of plasmid cDNA of each sample was normalized to the amount of protein produced in given cells in 48 h measured by the Western Blot signal intensity normalized to actin.

On 1st and 2nd days after transfection, the cells were trypsinized and manually counted. On day 3 after transfection, the cells were incubated with 10 µM BrdU (37 °C, 45min in complete media) and were collected for cell cycle assay and manual count. Then, the cells were fixed and stained with anti-BrdU antibodies and 7-AAD DNA dye according to FITC BrdU flow kit (BD Pharmingen™) manual instruction. The cells were excited at 488nm, and signals from 50,000 cells were acquired at 585/42 and 702/64 in LSR II (BD Biosciences). The results were analyzed using FlowJo software and were expressed as the percentage of cells in each cell cycle phase within the entire population excluding debris and apoptotic cells. At least triplicates were used for each sample and each experiment was conducted four times.

2.7. Western Blot and Subcellular Fractionation

To detect protein expression by Western Blotting, the cells were plated down in 6-well plates at a concentration 0.65×10^6 per well. The next day, the cells were transfected with plasmid cDNA using Lipofectamine™ 2000 (Thermo Scientific) according to the manufacturer's instructions. The cells were lysed in RIPA buffer containing (mM): 0.1%SDS, 50 mM Tris, pH 7.4, 150mM NaCl, 1mM EDTA, 1% Triton X-100, 1% sodium deoxycholate, 1 mM NaF, 200μM Na_3VO_4, and 1× Halt Protease and Phosphatase Inhibitor Cocktail (Thermo Scientific), and were ruptured by sonication before centrifugation at $10,000 \times g$ for 20 min at 4 °C. The cell lysates were normalized for total protein with BCA assay (Bio-Rad DC Protein Assay). The samples with 4× NuPage sample buffer (Thermo Scientific) supplemented with dithiothreitol (DTT, 400mM) were heated at 70 °C for 5 min, cooled to RT, and 100 μg per lane were separated by NuPAGE Bis-Tris 4–12% gradient gel (Thermo Scientific) in MES running buffer (Thermo Scientific). The gels were transferred in 10% Methanol-containing transfer buffer to FluoroTrans PVDF membranes (Pall), which were subsequently blocked in 5% non-fat milk (Carnation) in TNT buffer (0.1% Tween-20, 150 mM NaCl, 50mM Tris pH 8.0) for 1 h at RT. Membranes were probed overnight with primary antibodies diluted in 5% milk in TNT. Mouse monoclonal anti-Cx43 directed against C-terminal region, prepared to the last 23 amino acids of Cx43 (1:500, Millipore), mouse monoclonal anti-β-actin (1:2000, Sigma-Aldrich), mouse monoclonal anti-V5 (1:500, Sigma-Aldrich), and goat secondary antibodies conjugated to AlexaFluor 647 (1:500, Thermo Scientific) were used in this study. The membranes were imaged with a ChemiDocMP4000 fluorescent western detection system (BioRad). The membranes were stripped using Re-Blot plus Strong solution (Millipore) and re-probed with β-actin to ensure equal loading and were used as a normalization control. The relative intensity of signals was quantified using BioRad software.

The Nuclear Extraction kit (ab113474, Abcam) was used to obtain nuclei, and cytosolic-enriched fractions from HEK293FT cells transfected with GJA1-11k, GJA1-43k, cDNA plasmid, or negative control GFP-V5 cDNA. After reconstitution in RIPA buffer, fractions were analyzed by Western Blot in Bis-Tris 10% SDS-PAGE gel (100μg protein/lane), transferred to PVDF membrane and probe to mouse monoclonal anti-Cx43 directed against the C-terminal region (1:500, Millipore), mouse monoclonal anti-V5 (1:200, Sigma-Aldrich). To confirm the enrichment of correct proteins in obtained fractions, the same blot was probed with nuclear marker mouse monoclonal anti-Histone H3 (1:2000, Abcam), cytoplasmic marker mouse monoclonal anti-GAPDH (1:5000, Abcam), membrane marker mouse monoclonal NA+/K+-ATPase (1:10000, Millipore), Golgi marker rabbit monoclonal anti-GMP130 (1:1000, Abcam), and nuclear envelope marker rabbit monoclonal anti-Lamin B1 (1:5000, Abcam).

2.8. Statistical Analysis

All quantitative data are presented as mean ± SEM, with 'n' denoting of the number of independent experiments. The normality of the data sets was tested using Kolmogorov-Smirnov's test and d'Agostino and Pearson's test. For comparison between the two groups, an unpaired two-tail Student's *t*-test was performed. For comparison among three or more groups, one-way ANOVA (followed by Tukey's post-hoc test) or two-way ANOVA (followed by Tukey's multiple comparison test) were used accordingly and analyzed in Prism 6 software (GraphPad). A *p*-value < 0.05 was considered significant.

3. Results

3.1. Alternatively Translated Cx43 Isoforms Localize to Different Subcellular Compartments

We constructed separate cDNA plasmids, each expressing an alternatively translated isoform that can be generated from internal methionine start sites (GJA1-32k, GJA1-29k, GJA1-26k, GJA1-20k, GJA1-11k, GJA1-7k). For each of the plasmids, the AUGs downstream of each start codon were replaced by CUGs so that each plasmid only produces one Cx43 protein isoform fused with a C-terminal V5-tag. We also constructed a wild-type Cx43 (GJA1-WT) plasmid with all internal AUG start sites intact and a plasmid encoding only full-length Cx43 (GJA1-43k), which preserved the first start codon yet with each downstream AUG mutated to CUG (Figure S1). Human embryonic kidney HEK293FT cells have little endogenous Cx43 and therefore present little endogenous competition to the exogenous overexpression of Cx43 isoforms.

Using immunofluorescence, we used two separate antibodies (anti-Cx43 C-terminus and anti-V5) to identify the subcellular localization of each V5-tagged isoform when exogenously introduced in HEK293FT cells (Figure 1A). As expected, GJA1-WT localized as large plaque formations on the cell–cell border. In contrast, full length GJA1-43k accumulated in perinuclear regions, indicating a reduced ability to be trafficked to the cell border, consistent with our previously reported results of a six-fold reduction in gap junction plaques [22]. GJA1-20k, a chaperone that directs the trafficking of the full-length GJA1-43k isoform to the cell–cell border, generally appears localized to the perinuclear and ER regions and the mitochondria [26]. Of note, we did not detect significant GJA1-20k localization to the cell nucleus. However, GJA1-11k appeared to be highly enriched in the nucleus, with minimal detection in the cytoplasm.In general, V5 detection is diminished as the isoforms get smaller, yet the nucleus to cytoplasm ratios remained consistent. GJA1-7k, the smallest isoform, was poorly detected by either antibody. We also stained non-transfected cells to establish the baseline endogenous expression of Cx43 in HEK293FT cells.

A.

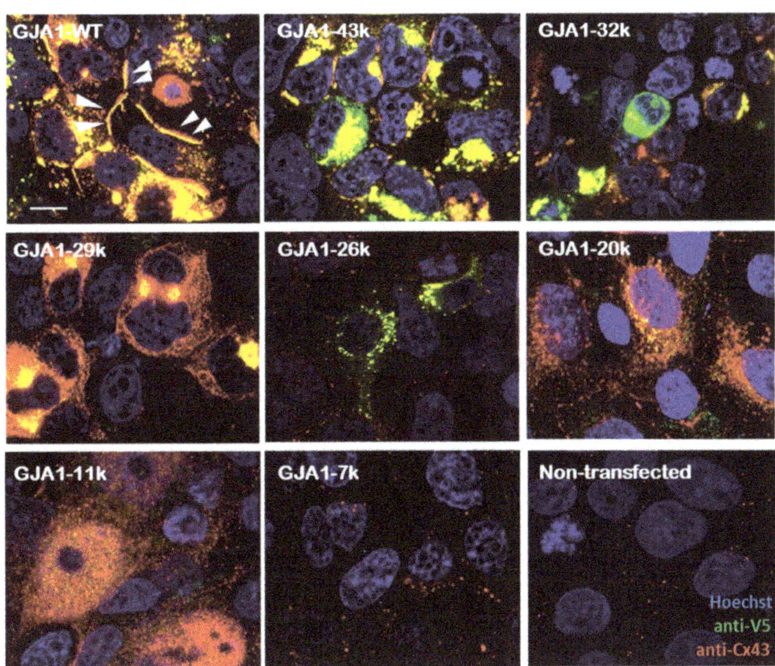

Figure 1. *Cont.*

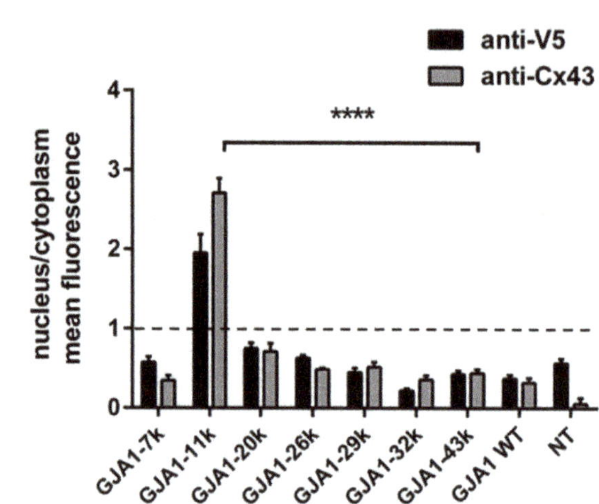

Figure 1. Localization of alternatively translated Cx43 isoforms expressed in HEK293FT cells. (**A**). The fixed-cell immunofluorescence of HEK293FT cells expressing short isoforms of Cx43 was detected by monoclonal anti-V5 (green) and polyclonal anti-Cx43 (red) antibodies, raised against a 17-residue peptide of C-terminal Cx43 (Sigma) with Hoechst staining of the nuclei (blue). The arrow heads show the gap junction (plaque) formation at the cell–cell border of cells with wild type (GJA1-WT) plasmid. Scale bar: 10μm. The results are representative of four independent experiments. (**B**). The ratio of fluorescent intensity in the nucleus versus cytoplasm in transfected HEK293FT cells. The data are presented as mean ± SEM, n = 20, **** $p < 0.0001$, by two-way ANOVA followed by Tukey's multiple comparison test. The intensities were measured and normalized to background using Image J.

To quantify the distribution of each isoform, we calculated the ratio of the mean fluorescent density in the nucleus versus cytoplasm (Figure 1B). This analysis confirmed that GJA1-11k has a two (anti-V5 antibody detection) to threefold (anti-Cx43 detection) nuclear enrichment. Furthermore, the nuclear localization is specific to GJA1-11k. All other isoforms capable of being generated by alternative translation had 50% or less nucleus to cytoplasm density. To confirm that nucleus localization was not secondary to the V5 tag interacting with GJA1-11k, we obtained similar results by a staining for HA-tagged GJA1-11k (Figure S2).

The nucleus localization of GJA1-11k was also tested by biochemistry techniques. We performed subcellular fractionations of HEK293FT cells overexpressing GJA1-11k, GJA1-43k, and a negative control GFP-V5. Consistent with the immunofluorescence data, Western Blot analysis of cytosolic and nuclear fractions revealed that when compared to full length GJA1-43k, GJA1-11k expression is more enriched in the nucleus, with nearly as much protein in the nucleus as in the cytoplasm despite a much smaller intranuclear volume (Figure 2, Figure S3). Together, the data of Figures 1 and 2 demonstrate that a major fraction of GJA1-11k, unlike the other mammalian Cx43 isoforms, localizes to the nucleus.

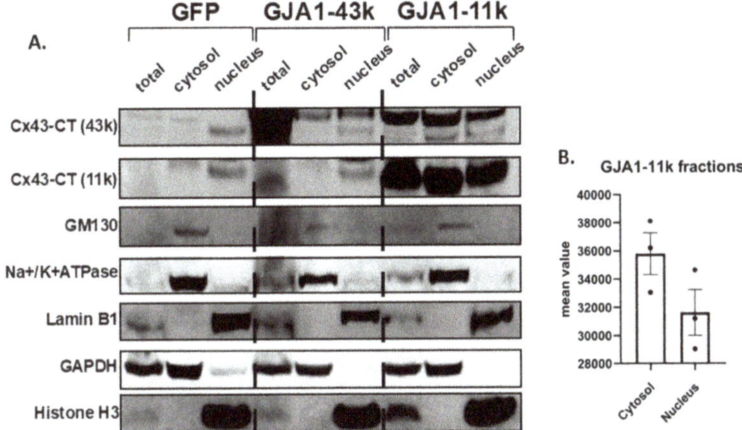

Figure 2. (A). Biochemical isolation of GJA1-11k in the nucleus. Western Blot of Cx43 isoforms in subcellular fractions (cytosolic and nuclear fractions) of HEK293FT cells over-expressing exogenous GJA1-11k, GJA1-43k, and GFP-V5 tagged plasmid cDNA, probed to monoclonal Cx43-CT (Millipore) antibodies. To demonstrate enriched biochemical isolation, the same blot was probed using antibodies to different subcellular fraction markers, including cytoplasmic (GAPDH), membrane (Na^+/K^+-ATPase), Golgi (GM-130), and nuclear markers (Histone H3, Lamin B1). The results are representative of three independent experiments. (B). The densitometry of CT-43k (11k) subcellular fractions assessed by Western Blot. Uncut immunoblots are provided in Figure S3.

3.2. GJA1-11k Inhibits Cell Proliferation

The nucleus is the primary organelle involved in cell cycle regulation. Because the C-terminus fragments are associated with inhibiting cell proliferation [29,31] and GJA1-11k appears to be enriched in the nucleus (Figures 1 and 2), we quantified cell proliferation in the presence of GJA1-11k. HEK293FT cells were transfected with the siRNA that silenced endogenous GJA1 mRNA, removing any endogenous source of any Cx43 isoform. Note that silencing GJA1 alone, even with little background signal, resulted in a significant increase in cell growth compared to non-transfected cells or scramble siRNA control (Figure 3, Figure S4). These data suggest that even minute quantities of Cx43 or its isoforms can inhibit cell proliferation.

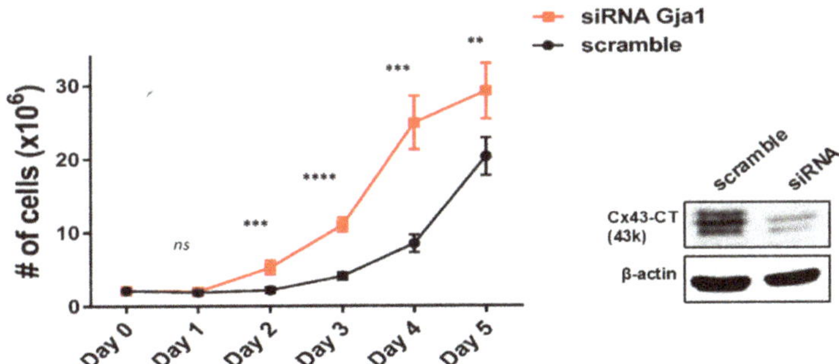

Figure 3. Effect of Cx43 on cell proliferation. The knockdown of endogenous Cx43 increases the cell proliferation of HEK293FT cells. For each day, data are presented as mean ± SEM, $n = 9$, ** $p < 0.01$, *** $p < 0.001$, **** $p < 0.0001$ by an unpaired two-tail Student's *t* test. Western Blot analysis confirmed the knockdown of Cx43 and is representative of three independent experiments.

We then explored the effect of the full length Cx43 and the GJA1-11k isoforms on cell proliferation. We transfected HEK293FT with vectors expressing GJA1-11k, GJA1-43k, GJA1-WT, and GFP-V5 as a negative control. The cells were counted over a three-day period (Figure 4). GJA1-11k significantly inhibited cell growth ($p < 0.01$), relative to GJA1-43k and GFP-V5. Apoptosis was ruled out by TUNEL staining (Figure S5). Together, these results indicate that GJA1-11k can be a potent growth suppressor, more so than either the full-length isoform GJA1-43k or the intact WT Cx43.

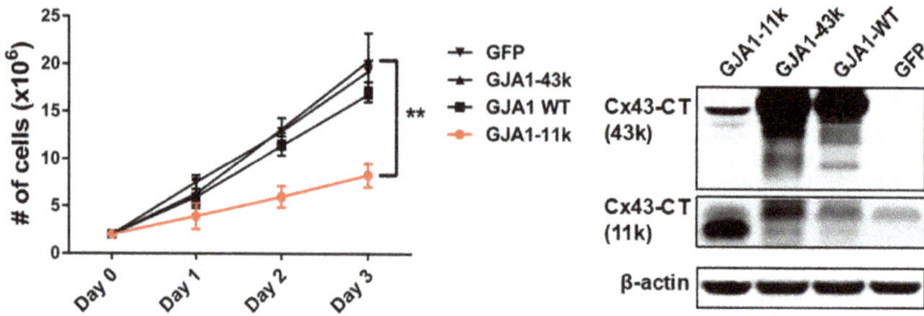

Figure 4. Effect of Cx43 short isoforms on cell proliferation. The overexpression of GJA1-11k inhibits the cell growth of the HEK293FT cell line. The growth curve is counts of transiently transfected cells each with vectors expressing GJA1-11k, GJA1-43k, GJA1-WT, and the negative control plasmid GFP with V5-tag. For each day data are presented as mean ± SEM, $n = 12$, ** $p < 0.01$ by one-way ANOVA followed by Tukey's post-hoc test. Cell growth analysis revealed that cells expressing GJA1-11k grew significantly slower, than cells overexpressing other isoforms. Western Blot confirmed transient expression of Cx43 isoforms. Uncut immunoblots are provided in Figure S6.

We next tested whether nuclear localization is essential for GJA1-11k to convey growth inhibition. From an in silico analysis, we could not find a nuclear localization sequence (NLS) in GJA1-11k to remove and therefore limit its entry. However, we were able to add a nuclear export signal (NES) [33] to GJA1-11k. A vector was generated with an NES sequence (LQLPPLERLTLD) [34] added to GJA1-11k-V5, and transiently transfected into HEK293FT. We then used immunofluorescence to identify changes in nuclear localization. The expression of GJA1-11k-NES significantly reduced the presence of GJA1-11k in the nucleus (Figure 5A), $p < 0.0001$, whether detection was by antibodies against Cx43-CT or V5 tag (Figure 5A, right Panel), $p < 0.001$. We next quantified cell growth as done in Figure 4. The results show that GJA1-11k-NES partially rescues cell growth (Figure 5B, Figure S6). Residual decreases in cell count were likely due to residual GJA1-11k remaining in the nucleus.

3.3. GJA1-11k Inhibits Cell Cycle Progression to the S Phase

The inhibition of proliferation implies interruption of cell cycle progression. We explored the stage at which cell growth is inhibited by GJA1-11k in the nucleus. HEK293FT cells were serum-deprived for 48h to synchronize cells in the phase G_0/G_1 prior to transfection. Seventy-two hours after transfection, the cells were labeled with BrdU for 45 min and subsequently fixed and stained with anti-BrdU antibody (protocol diagrammed in Figure S7). Flow cytometry was used to quantify the percentage of cells in each phase (Figure 6A). Quantification was performed in cells transfected with either a negative control (GW-CAT V5), GJA1-WT, full length GJA1-43k, GJA1-11k, or GJA1-11k-NES. Our results in Figure 6B indicate that, of all the plasmids, GJA1-11k resulted in the highest percentage of cells remaining in G0/G1 (from 17% in the negative control to almost 40% with GJA1-11k, a twofold increase). As would be expected by capturing cells in G0/G1, the expression of GJA1-11k results in half of the cells in the S phase as it does in the negative control cohort (from 50% to 27%, $p < 0.05$; Figure 6B).

The expression of GJA1-11k-NES resulted in increasing the percentage of cells in the S phase relative to GJA1-11k ($p < 0.01$; Figure 6B).

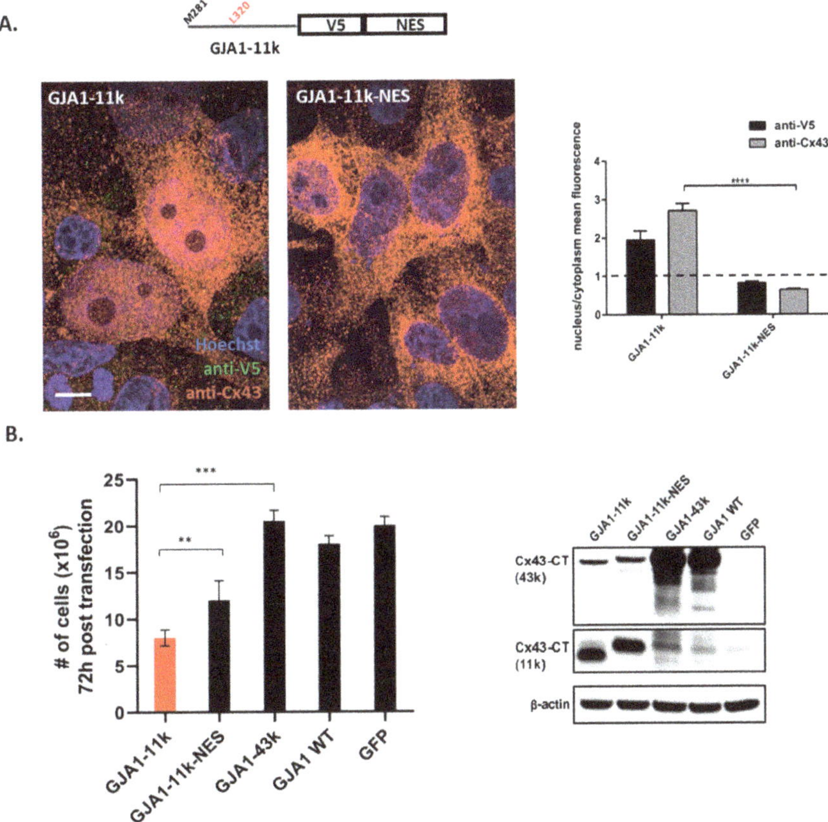

Figure 5. GJA1-11k with nuclear export signal rescues the cell proliferation of the HEK293FT cell line. (**A**). The immunofluorescence of HEK293FT cells expressing GJA1-11k or GJA1-11k-NES probed with V5 and Cx43 antibodies (Sigma) and the corresponding ratio of fluorescence density in the nucleus versus cytoplasm is shown on the graph. Scale bar: 10μm. The graph results are representative of four independent experiments. The data are presented as mean ± SEM, n = 26, **** $p < 0.0001$ by an unpaired two-tail Student's *t* test. The intensities in the bar graph were measured and normalized to background using Image J. (**B**). GJA1-11k-NES rescues cell proliferation, suggesting that the effect of GJA1-11k is linked to the nucleus. The number of cells transiently transfected with vectors expressing GJA1-11k, GJA1-11k-NES, GJA1-43k, GJA1-WT, and the negative control plasmid GFP were counted 72 h after transfection. Cell growth analysis has shown that cells expressing GJA1-11k-NES grew significantly faster than cells expressing GJA1-11k. The Western Blot confirmed the transient expression of Cx43 isoforms. The data are presented as mean ± SEM, *n* = 12, ** $p < 0.01$, *** $p < 0.001$ by one-way ANOVA followed by Tukey's post-hoc test.

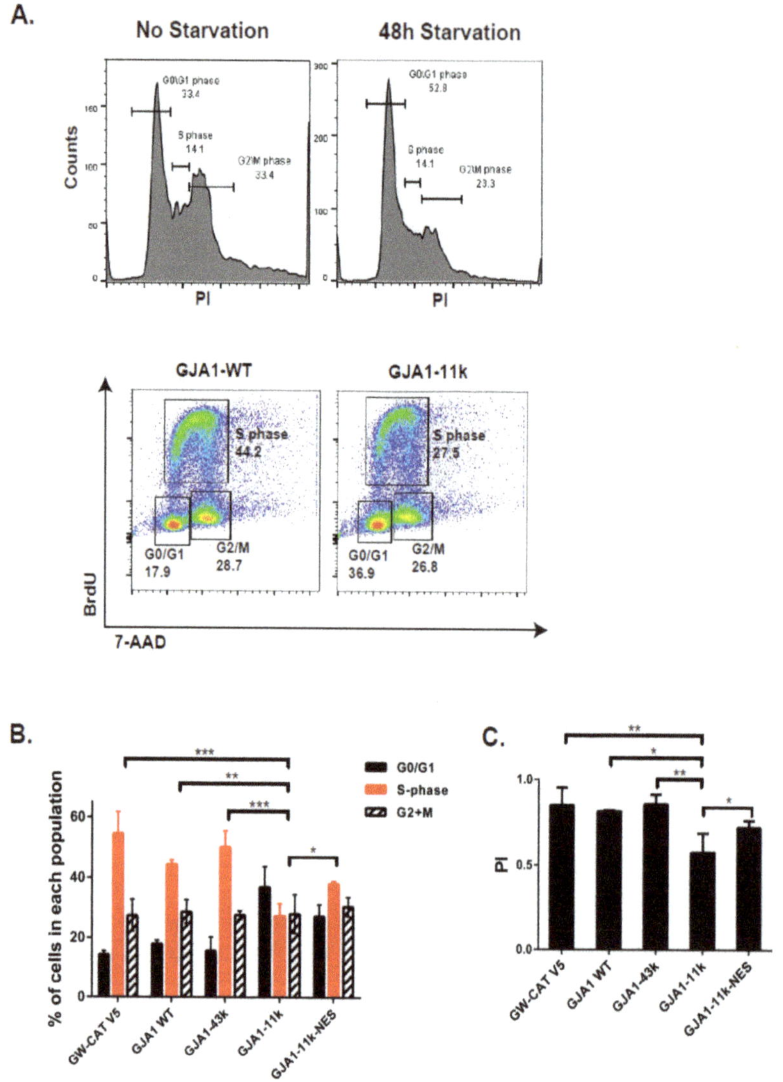

Figure 6. Overexpression of wild type Cx43 and short isoforms inhibits cell cycle progression in HEK293FT cells. (**A**). Non-transfected cells were serum-deprived for 48 h to synchronize cells in the G_0/G_1 phase. A representative image showing cell cycle phases in the HEK293FT cell line based on BrdU incorporation versus total DNA staining by 7-AAD. The cell cycle phases were analyzed by FlowJo. (**B**). The cell cycle distribution of GJA1-WT, GJA1-43k, GJA1-11k, and GJA1-11k NES V5-tagged isoforms were analyzed 72 h post-transfection. The bar graph shows the reduction in the number of cells in the S-phase and accumulation of cells in G_0/G_1 expressing GJA1-11k compare to GJA1-WT or GJA1-43k. GJA1-11k-NES rescues cell cycle progression from G_0/G_1 to S-phase. The data are presented as mean ± SEM, $n = 12$, * $p < 0.05$ and ** $p < 0.01$, *** $p < 0.001$ by one-way ANOVA followed by Tukey's post-hoc test. (**C**). The Proliferation index (using following PI = ($(S+G_2/M) / (G_0/G_1+S+G_2/M)$) formula) confirmed the reduction in proliferation rate for GJA1-11k. The data are presented as mean ± SEM, $n = 12$, * $p < 0.05$ and ** $p < 0.01$, by one-way ANOVA followed by Tukey's post-hoc test.

The flow cytometry data in Figure 6B can be used to quantify growth fraction by a proliferation index (PI). Of all the plasmids, only GJA1-11k significantly reduced the calculated PI (Figure 6C). Notably, the addition of the NES to GJA1-11k reverses the GJA1-11k induced reduction in PI (Figure 6C). Together, the results confirm a GJA1-11k induced reduction in the proliferation that occurs by limiting the exit of the G0/G1 phase and that nucleus localization is necessary to affect the growth inhibition.

3.4. Discussion

Our work attributes a specific biological role to mammalian GJA1-11k. The 11kDa alternatively translated isoform of Cx43 functions independent of gap junction formation as it uniquely localizes to the nucleus where it directly affects cell growth, limiting cell cycle progression from the G0/G1 phase to the S phase.

In our expression analysis (Figure 1), we found mammalian GJA1-20k to be distributed in cytoplasmic organelles, consistent with its known roles in Cx43 forward trafficking [22,24] and metabolism [26,27]. This reported localization differs somewhat to a recently published study about the role of Xenopus GJA1-20k, which was found to translocate to the nucleus of Xenopus and HeLa cells (where it can upregulate N-cadherin expression [28]). It is not clear why we could not reproduce Kotini et al's [28] finding regarding GJA1-20k nucleus enrichment. The Xenopus GJA1-20k has a different sequence and is shorter than human GJA1-20k by six amino acids. The Xenopus plasmid also did not mutate downstream methionine resulting in potential generation of alternative isoforms GJA1-11kDa and 7kDa, which may have been detected in lieu of GJA1-20k. The Xenopus GJA1-20k in the Kotini study [28] was also covalently bound to the glucocorticoid receptor to enforce nuclear translocation of GJA1-20k, and thus nuclear localization could have been more a product of the glucocorticoid receptor that of the GJA1-20k itself. Finally, differences between the human and Xenopus GJA1-20k isoform sequences could result in differences in localization tendencies. Future studies carefully exploiting the differences in the plasmids and their effect on localization should be highly informative.

Cx43 expression is typically reduced in tumor cell lines [17] or tissues [8,18–20,35–38] and the loss of Cx43 is associated with shorter patient survival [8,20]. It has already been identified that ectopic expression of WT Cx43 in cancer cells does not result in gap junctions generation yet limits cell proliferation [7,8,39,40]. Despite these observations and correlations, significant knowledge gaps exist in our mechanistic understanding of non-canonical effects of ectopic Cx43. It should also be noted that most studies identify Cx43 by antibody staining in which the epitope for Cx43 specific antibodies is in the C-terminus; therefore, full length Cx43 cannot be differentiated from the smaller isoforms [22]. We expect that reports of Cx43 in the nucleus in cancer cells [28–31] indeed may be GJA-11k, which is protective against cancer progression. For exogenous introduction, GJA1-11k is a small, hydrophilic, nucleus-targeted peptide that could be exploited for cancer therapy. Future studies should explore the molecular mechanisms underlying nuclear, GJA1-11k-regulated cell cycle progression and cell growth, as well as its translation into a new therapy targeting tumor growth and progression. In this particular study, we used an in vitro model with only HEK cells which have fast growth. In addition, we relied on exogenous expression that can result in super-physiologic levels of transfected proteins. Finally, HEK cells have limitations in microscopy that can be overcome with flatter cells. Future studies should include multiple cell lines including cancer cell lines, as well as in vivo models.

Given the identified roles of GJA1-20k in channel trafficking [22,24], metabolism [25–27,41], transcription [28], as well as in epithelial-mesenchymal transition [42], we were surprised in this study that the dominant phenotype of GJA1-11k relates to yet another phenomenon, specifically inhibiting cell cycle progression. The common denominator, we believe, is the affinity of the Cx43 C-terminus for both microtubule [26] and actin [24] cytoskeletons. A microtubule-binding domain for Cx43 has been identified in the proximal Cx43 C-terminus [43,44]. If actin is involved in GJA1-11k nuclear localization, then the results of the current study would support early studies identifying an association between Cx43 and actin [24,45] and, in general, help us understand that cytoskeletal interactions are the basis for multiple fundamental properties of the internally translated isoforms of Cx43.

Our focus on cell cycle regulation was based on a combination of the classic role of proteins with intranuclear enrichment and prior reports that Cx43 involvement in cell-cycle regulation is not easily explained by sarcolemmal localized ion channels. As the field explores the effects of Cx43's smaller isoforms, we expect the number of cellular phenomena associated with each isoform to grow and we will identify crossover between the roles of different isoforms. For instance, in unpublished studies, we observed that GJA1-11k can affect the transcription of GJA1 mRNA. Already, GJA1-20k is associated with the transcriptional regulation of N-Cadherin [28]. Similarly, proteins that rearrange the extranuclear cytoskeleton, such as GJA1-20k [24], may also affect the mechanics of cell proliferation. We look forward to the next several years of discovering the roles of Cx43's smaller isoforms.

Formation of smaller truncated isoforms by alternative translation can be a source of considerable biologic diversity for any gene, and especially single coding exon genes such as Gja1 that are not able to undergo splicing. Traditionally, alternative translation is considered a phenomenon that occurs with development and evolutionary changes [46]. However, it has already been established that the expression of GJA1-20k and other smaller isoforms increase with metabolic stress [25,27] or as a result of mTOR and Mnk1/2 pathway inhibition. These pathways are involved in critical and dynamic cellular processes, such as cell growth, proliferation, transcription, and survival [22,23], and occur in mature cells on a much more acute time scale than during development and evolutionary change. Previous findings have identified GJA1-20k as a protective stress response protein in terminally differentiated cardiomyocytes [27]. Based on our results, GJA1-11k could also be a stress response protein, but its upregulation could be a response to inhibit cell cycle progression.

Supplementary Materials: The following are available online at http://www.mdpi.com/2218-273X/10/3/473/s1, Figure S1: title, Table S1: title, Video S1: title.

Author Contributions: Conceptualization, I.E., S.X., T.T.H. and R.M.S.; Methodology, I.E., S.X.; Investigation, I.E.; Writing—Original Draft, I.E.; Writing—Review & Editing, I.E., R.E.B., A.G.K. and R.M.S.; Funding Acquisition, A.K., T.T.H. and R.S.; Supervision, S.X., A.G.K., T.T.H. and R.M.S. All authors have read and agreed to the published version of the manuscript.

Funding: This work was supported by National Institute of Health grants HL136463 (A.K.), HL133286 (T.T.H.), HL138577 and HL152691 (R.M.S.).

Acknowledgments: We thank Flow core at Cedars-Sinai Medical Center for technical assistance.

Conflicts of Interest: We have no conflict of interest to disclose.

References

1. Goodenough, D.A.; Goliger, J.A.; Paul, D.L. Connexins, connexons, and intercellular communication. *Annu. Rev. Biochem.* **1996**, *65*, 475–502. [CrossRef]

2. Beyer, E.C.; Paul, D.L.; Goodenough, D.A. Connexin43: A protein from rat heart homologous to a gap junction protein from liver. *J. Cell Biol.* **1987**, *105*, 2621–2629. [CrossRef] [PubMed]

3. Shaw, R.M.; Rudy, Y. Ionic mechanisms of propagation in cardiac tissue. Roles of the sodium and L-type calcium currents during reduced excitability and decreased gap junction coupling. *Circ. Res.* **1997**, *81*, 727–741. [CrossRef] [PubMed]

4. Beardslee, M.A.; Lerner, D.L.; Tadros, P.N.; Laing, J.G.; Beyer, E.; Yamada, K.A.; Kléber, A.G.; Schuessler, R.B.; Saffitz, J.E. Dephosphorylation and intracellular redistribution of ventricular connexin43 during electrical uncoupling induced by ischemia. *Circ. Res.* **2000**, *87*, 656–662. [CrossRef] [PubMed]

5. Dupont, E.; Matsushita, T.; Kaba, R.A.; Vozzi, C.; Coppen, S.R.; Khan, N.; Kaprielian, R.; Yacoub, M.H.; Severs, N.J. Altered connexin expression in human congestive heart failure. *J. Mol. Cell. Cardiol.* **2001**, *33*, 359–371. [CrossRef]

6. Aasen, T.; Mesnil, M.; Naus, C.C.; Lampe, P.D.; Laird, D. Gap junctions and cancer: Communicating for 50 years. *Nat. Rev. Cancer* **2016**, *16*, 775–788. [CrossRef]

7. Kardami, E.; Dang, X.; Iacobas, D.A.; Nickel, B.E.; Jeyaraman, M.; Srisakuldee, W.; Makazan, J.; Tanguy, S.; Spray, D.C. The role of connexins in controlling cell growth and gene expression. *Prog. Biophys. Mol. Biol.* **2007**, *94*, 245–264. [CrossRef]

8.   Sirnes, S.; Bruun, J.; Kolberg, M.; Kjenseth, A.; Lind, G.E.; Svindland, A.; Brech, A.; Nesbakken, A.; Lothe, R.A.; Leithe, E.; et al. Connexin43 acts as a colorectal cancer tumor suppressor and predicts disease outcome. *Int. J. Cancer* **2012**, *131*, 570–581. [CrossRef]

9.   Mesnil, M.; Aasen, T.; Boucher, J.; Chepied, A.; Cronier, L.; Defamie, N.; Kameritsch, P.; Laird, D.W.; Lampe, P.D.; Lathia, J.D.; et al. An update on minding the gap in cancer. *Biochim. Et Biophys. Acta* **2018**, *1860*, 237–243. [CrossRef]

10.   Becker, D.L.; Thrasivoulou, C.; Phillips, A.R. Connexins in wound healing; perspectives in diabetic patients. *Biochim. Et Biophys. Acta* **2012**, *1818*, 2068–2075. [CrossRef]

11.   Araya, R.; Eckardt, D.; Maxeiner, S.; Kruger, O.; Theis, M.; Willecke, K.; Saez, J.C. Expression of connexins during differentiation and regeneration of skeletal muscle: Functional relevance of connexin43. *J. Cell Sci.* **2005**, *118*, 27–37. [CrossRef] [PubMed]

12.   Liu, X.; Sun, L.; Torii, M.; Rakic, P. Connexin 43 controls the multipolar phase of neuronal migration to the cerebral cortex. *Proc. Natl. Acad. Sci. USA* **2012**, *109*, 8280–8285. [CrossRef] [PubMed]

13.   Mathias, R.T.; White, T.W.; Gong, X. Lens gap junctions in growth, differentiation, and homeostasis. *Physiol. Rev.* **2010**, *90*, 179–206. [CrossRef] [PubMed]

14.   Cina, C.; Maass, K.; Theis, M.; Willecke, K.; Bechberger, J.F.; Naus, C.C. Involvement of the cytoplasmic C-terminal domain of connexin43 in neuronal migration. *J. Neurosci. Off. J. Soc. Neurosci.* **2009**, *29*, 2009–2021. [CrossRef] [PubMed]

15.   Tittarelli, A.; Guerrero, I.; Tempio, F.; Gleisner, M.A.; Avalos, I.; Sabanegh, S.; Ortiz, C.; Michea, L.; Lopez, M.N.; Mendoza-Naranjo, A.; et al. Overexpression of connexin 43 reduces melanoma proliferative and metastatic capacity. *Br. J. Cancer* **2015**, *113*, 259–267. [CrossRef]

16.   Shao, Q.; Wang, H.; McLachlan, E.; Veitch, G.I.; Laird, D.W. Down-regulation of Cx43 by retroviral delivery of small interfering RNA promotes an aggressive breast cancer cell phenotype. *Cancer Res.* **2005**, *65*, 2705–2711. [CrossRef]

17.   Sin, W.C.; Crespin, S.; Mesnil, M. Opposing roles of connexin43 in glioma progression. *Biochim. Et Biophys. Acta* **2012**, *1818*, 2058–2067. [CrossRef]

18.   Laird, D.W.; Fistouris, P.; Batist, G.; Alpert, L.; Huynh, H.T.; Carystinos, G.D.; Alaoui-Jamali, M.A. Deficiency of connexin43 gap junctions is an independent marker for breast tumors. *Cancer Res.* **1999**, *59*, 4104–4110.

19.   Ableser, M.J.; Penuela, S.; Lee, J.; Shao, Q.; Laird, D.W. Connexin43 reduces melanoma growth within a keratinocyte microenvironment and during tumorigenesis in vivo. *J. Biol. Chem.* **2014**, *289*, 1592–1603. [CrossRef]

20.   Teleki, I.; Szasz, A.M.; Maros, M.E.; Gyorffy, B.; Kulka, J.; Meggyeshazi, N.; Kiszner, G.; Balla, P.; Samu, A.; Krenacs, T. Correlations of differentially expressed gap junction connexins Cx26, Cx30, Cx32, Cx43 and Cx46 with breast cancer progression and prognosis. *PLoS ONE* **2014**, *9*, e112541. [CrossRef]

21.   Fishman, G.I.; Eddy, R.L.; Shows, T.B.; Rosenthal, L.; Leinwand, L.A. The human connexin gene family of gap junction proteins: Distinct chromosomal locations but similar structures. *Genomics* **1991**, *10*, 250–256. [CrossRef]

22.   Smyth, J.W.S.; Shaw, R.M. Autoregulation of connexin43 gap junction formation by internally translated isoforms. *Cell Rep.* **2013**, *5*, 611–618. [CrossRef] [PubMed]

23.   Salat-Canela, C.; Sese, M.; Peula, C.; Ramon y Cajal, S.; Aasen, T. Internal translation of the connexin 43 transcript. *Cell Commun. Signal. Ccs* **2014**, *12*, 31. [CrossRef]

24.   Basheer, W.A.; Xiao, S.; Epifantseva, I.; Fu, Y.; Kleber, A.G.; Hong, T.; Shaw, R.M. GJA1-20k Arranges Actin to Guide Cx43 Delivery to Cardiac Intercalated Discs. *Circ. Res.* **2017**, *121*, 1069–1080. [CrossRef] [PubMed]

25.   Ul-Hussain, M.; Olk, S.; Schoenebeck, B.; Wasielewski, B.; Meier, C.; Prochnow, N.; May, C.; Galozzi, S.; Marcus, K.; Zoidl, G.; et al. Internal ribosomal entry site (IRES) activity generates endogenous carboxyl-terminal domains of Cx43 and is responsive to hypoxic conditions. *J. Biol. Chem.* **2014**, *289*, 20979–20990. [CrossRef] [PubMed]

26.   Fu, Y.; Zhang, S.S.; Xiao, S.; Basheer, W.A.; Baum, R.; Epifantseva, I.; Hong, T.; Shaw, R.M. Cx43 Isoform GJA1-20k Promotes Microtubule Dependent Mitochondrial Transport. *Front. Physiol.* **2017**, *8*, 905. [CrossRef] [PubMed]

27.   Basheer, W.A.; Fu, Y.; Shimura, D.; Xiao, S.; Agvanian, S.; Hernandez, D.M.; Hitzeman, T.C.; Hong, T.; Shaw, R.M. Stress response protein GJA1-20k promotes mitochondrial biogenesis, metabolic quiescence, and cardioprotection against ischemia/reperfusion injury. *JCI Insight* **2018**, *3*, e121900. [CrossRef]

28. Kotini, M.; Barriga, E.H.; Leslie, J.; Gentzel, M.; Rauschenberger, V.; Schambony, A.; Mayor, R. Gap junction protein Connexin-43 is a direct transcriptional regulator of N-cadherin in vivo. *Nat. Commun.* **2018**, *9*, 3846. [CrossRef]

29. Dang, X.; Doble, B.W.; Kardami, E. The carboxy-tail of connexin-43 localizes to the nucleus and inhibits cell growth. *Mol. Cell. Biochem.* **2003**, *242*, 35–38. [CrossRef]

30. Hatakeyama, T.; Dai, P.; Harada, Y.; Hino, H.; Tsukahara, F.; Maru, Y.; Otsuji, E.; Takamatsu, T. Connexin43 functions as a novel interacting partner of heat shock cognate protein 70. *Sci. Rep.* **2013**, *3*, 2719. [CrossRef]

31. Moorby, C.; Patel, M. Dual functions for connexins: Cx43 regulates growth independently of gap junction formation. *Exp. Cell Res.* **2001**, *271*, 238–248. [CrossRef] [PubMed]

32. Hino, H.; Dai, P.; Yoshida, T.; Hatakeyama, T.; Harada, Y.; Otsuji, E.; Okuda, T.; Takamatsu, T. Interaction of Cx43 with Hsc70 regulates G1/S transition through CDK inhibitor p27. *Sci. Rep.* **2015**, *5*, 15365. [CrossRef] [PubMed]

33. Lam, S.S.; Martell, J.D.; Kamer, K.J.; Deerinck, T.J.; Ellisman, M.H.; Mootha, V.K.; Ting, A.Y. Directed evolution of APEX2 for electron microscopy and proximity labeling. *Nat. Methods* **2015**, *12*, 51–54. [CrossRef] [PubMed]

34. Kosugi, S.; Hasebe, M.; Tomita, M.; Yanagawa, H. Nuclear export signal consensus sequences defined using a localization-based yeast selection system. *Traffic* **2008**, *9*, 2053–2062. [CrossRef]

35. Cronier, L.; Crespin, S.; Strale, P.O.; Defamie, N.; Mesnil, M. Gap junctions and cancer: New functions for an old story. *Antioxid. Redox Signal.* **2009**, *11*, 323–338. [CrossRef]

36. Ismail, R.; Rashid, R.; Andrabi, K.; Parray, F.Q.; Besina, S.; Shah, M.A.; Hussain, M.U. Pathological Implications of Cx43 Down-regulation in Human Colon Cancer. *Asian Pac. J. Cancer Prev.* **2014**, *15*, 2987–2991. [CrossRef]

37. Corteggio, A.; Florio, J.; Roperto, F.; Borzacchiello, G. Expression of gap junction protein connexin 43 in bovine urinary bladder tumours. *J. Comp. Pathol.* **2011**, *144*, 86–90. [CrossRef]

38. Talbot, J.; Brion, R.; Picarda, G.; Amiaud, J.; Chesneau, J.; Bougras, G.; Stresing, V.; Tirode, F.; Heymann, D.; Redini, F.; et al. Loss of connexin43 expression in Ewing's sarcoma cells favors the development of the primary tumor and the associated bone osteolysis. *Biochim. Et Biophys. Acta* **2013**, *1832*, 553–564. [CrossRef]

39. Sun, Y.; Zhao, X.; Yao, Y.; Qi, X.; Yuan, Y.; Hu, Y. Connexin 43 interacts with Bax to regulate apoptosis of pancreatic cancer through a gap junction-independent pathway. *Int. J. Oncol.* **2012**, *41*, 941–948. [CrossRef]

40. Mennecier, G.; Derangeon, M.; Coronas, V.; Herve, J.C.; Mesnil, M. Aberrant expression and localization of connexin43 and connexin30 in a rat glioma cell line. *Mol. Carcinog.* **2008**, *47*, 391–401. [CrossRef] [PubMed]

41. Wang, M.; Smith, K.; Yu, Q.; Miller, C.; Singh, K.; Sen, C.K. Mitochondrial connexin 43 in sex-dependent myocardial responses and estrogen-mediated cardiac protection following acute ischemia/reperfusion injury. *Basic Res. Cardiol.* **2019**, *115*, 1. [CrossRef] [PubMed]

42. James, C.C.; Zeitz, M.J.; Calhoun, P.J.; Lamouille, S.; Smyth, J.W. Altered translation initiation of Gja1 limits gap junction formation during epithelial-mesenchymal transition. *Mol. Boil. Cell* **2018**, *29*, 797–808. [CrossRef] [PubMed]

43. Saidi Brikci-Nigassa, A.; Clement, M.J.; Ha-Duong, T.; Adjadj, E.; Ziani, L.; Pastre, D.; Curmi, P.A.; Savarin, P. Phosphorylation controls the interaction of the connexin43 C-terminal domain with tubulin and microtubules. *Biochemistry* **2012**, *51*, 4331–4342. [CrossRef]

44. Giepmans, B.N.; Verlaan, I.; Hengeveld, T.; Janssen, H.; Calafat, J.; Falk, M.M.; Moolenaar, W.H. Gap junction protein connexin-43 interacts directly with microtubules. *Curr. Biol.* **2001**, *11*, 1364–1368. [CrossRef]

45. Smyth, J.W.; Vogan, J.M.; Buch, P.J.; Zhang, S.S.; Fong, T.S.; Hong, T.T.; Shaw, R.M. Actin cytoskeleton rest stops regulate anterograde traffic of connexin 43 vesicles to the plasma membrane. *Circ. Res.* **2012**, *110*, 978–989. [CrossRef] [PubMed]

46. de Klerk, E.; 't Hoen, P.A.C. Alternative mRNA transcription, processing, and translation: Insights from RNA sequencing. *Trends Genet.* **2015**, *31*, 128–139. [CrossRef]

Review

Trafficking and Function of the Voltage-Gated Sodium Channel β2 Subunit

Eric Cortada [1,2], Ramon Brugada [1,2,3,4] and Marcel Verges [1,2,3,*]

[1] Cardiovascular Genetics Group, Girona Biomedical Research Institute (IDIBGI), C/ Doctor Castany, s/n—Edifici IDIBGI, 17190 Girona, Spain; ecortada@gencardio.com (E.C.); rbrugada@idibgi.org (R.B.)
[2] Biomedical Research Networking Center on Cardiovascular Diseases (CIBERCV), 28029 Madrid, Spain
[3] Medical Sciences Department, University of Girona Medical School, 17003 Girona, Spain
[4] Cardiology Department, Hospital Josep Trueta, 17007 Girona, Spain
* Correspondence: mverges@gencardio.com; Tel.: +34-872-987-087 (ext. 62)

Received: 17 September 2019; Accepted: 8 October 2019; Published: 13 October 2019

Abstract: The voltage-gated sodium channel is vital for cardiomyocyte function, and consists of a protein complex containing a pore-forming α subunit and two associated β subunits. A fundamental, yet unsolved, question is to define the precise function of β subunits. While their location in vivo remains unclear, large evidence shows that they regulate localization of α and the biophysical properties of the channel. The current data support that one of these subunits, β2, promotes cell surface expression of α. The main α isoform in an adult heart is $Na_V1.5$, and mutations in *SCN5A*, the gene encoding $Na_V1.5$, often lead to hereditary arrhythmias and sudden death. The association of β2 with cardiac arrhythmias has also been described, which could be due to alterations in trafficking, anchoring, and localization of $Na_V1.5$ at the cardiomyocyte surface. Here, we will discuss research dealing with mechanisms that regulate β2 trafficking, and how β2 could be pivotal for the correct localization of $Na_V1.5$, which influences cellular excitability and electrical coupling of the heart. Moreover, β2 may have yet to be discovered roles on cell adhesion and signaling, implying that diverse defects leading to human disease may arise due to β2 mutations.

Keywords: cardiac arrhythmias; protein trafficking; voltage-gated sodium channel; $Na_V1.5$; *SCN2B*

1. Introduction

We often wonder why people get sick. Why do some people die early in life while others reach old age without major health problems? It is obvious that both genetic and environmental factors are playing a role. An organ that must certainly remain in perfect condition throughout life is the heart; its abnormal functioning is a sign of diseases that can lead to premature death. An unsolved matter is to determine the molecular alterations that can lead to heart disease and how these may arise.

1.1. Arrhythmias and Sudden Cardiac Death (SCD)

Cardiomyopathies are a type of disease affecting the function of the heart muscle and leading to heart failure. Suffering from cardiomyopathy also implies being at risk of arrhythmia and sudden cardiac death (SCD). In the US alone, SCD affects annually up to 400,000 people, with coronary heart disease being present in 80% of the cases [1]. Importantly, it is estimated that around 6 million people die each year worldwide of SCD due to ventricular arrhythmia, which is often associated with cardiomyopathy (the survival rate of a cardiac arrest is <1%). This represents 1/3rd of deaths from cardiovascular disease of any kind, being these in fact the main cause of death in the world [2]. Therefore, understanding how arrhythmias develop, and the consequences that they entail, has interest in the fields of biology, medicine and socioeconomics.

SCD can occur in people under 50 and be associated with pathologies defined as unexplained or arrhythmogenic. It is often the first and single clinical manifestation of an inherited heart disease that has remained undiagnosed by conventional clinical practice [3]. A well-known arrhythmia is Brugada syndrome (BrS), a disorder characterized by an abnormal electrocardiogram (ECG) that causes ventricular fibrillation. Although considered a rare genetic disease, it is actually inherited by an autosomal dominant trait [4] and linked to a high mortality rate [5]. In BrS, the ventricular muscle quivers, instead of contracting in a coordinated manner. Such tachycardia prevents the blood from flowing efficiently throughout the body. Consequently, the individual faints and can die within a few minutes [6].

Since genetics and molecular biology emerged in the field of clinical cardiology, multiple genes and mutated variants responsible for arrhythmias that lead to SCD were identified. A fundamental hypothesis is that many cases of SCD are the clinical manifestation of rare hereditary diseases caused by mutations in transmembrane ion channels implicated in generating the cardiac action potential (AP). These channelopathies are due to alterations in subunits of the channel protein complex, as well as in associated or regulatory proteins. Therefore, they can be congenital and due to one or more mutations in the genes involved. Along with BrS, other channelopathies that cause SCD include long QT syndrome (LQTS) and catecholaminergic polymorphic ventricular tachycardia [7].

1.2. The Cardiac Voltage-Gated Sodium (Na$_V$) Channel

The voltage-gated sodium (Na$_V$) channel often shows alterations leading to cardiac channelopathies [8]. For instance, ~20 % of BrS cases are caused by mutations in *SCN5A*, a gene mapping to the chromosomal region 3p21 [9] and encoding Na$_V$1.5, the pore-forming, α subunit, of the main cardiac Na$_V$ channel [4]. In the heart, the Na$_V$ channel is responsible for generating the rising phase of the AP, thus playing a central role in myocardial excitability. The abnormal ECG observed in BrS is due to loss-of-function of the Na$_V$ channel. It is characterized by an ST segment elevation of the V1-V3 precordial leads and occurs in the absence of structural heart disease [6] (Figure 1).

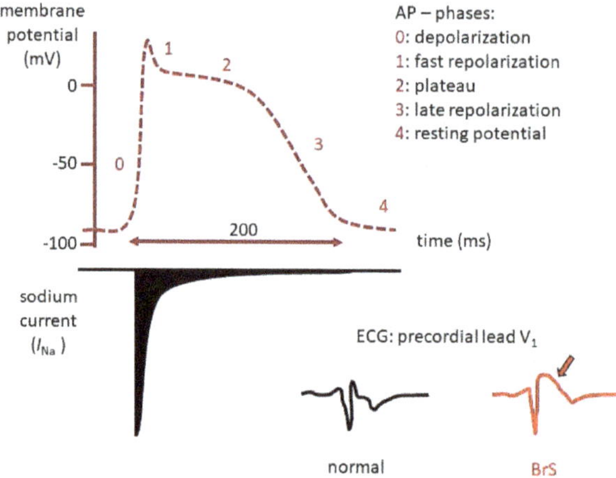

Figure 1. Role of Na$_V$1.5 in the generation of the cardiac action potential (AP). Schematic representation of the cardiac AP and the contribution of the sodium current (I_{Na}) generated by Na$_V$1.5 as responsible for the rising (depolarization) phase. The electrical pattern of precordial lead V1 of a normal electrocardiogram (ECG) and one with an ST segment elevation typical of BrS are shown (arrow).

Initiation of the AP, along with the degree of intercellular communication via gap junctions, to allow propagation, determines the conduction velocity of the electrical impulse. Thus, mutations in

SCN5A have also been associated with LQTS, atrial fibrillation and even cardiomyopathies [10]. In fact, the Na_V channel also plays an important role in electrical impulse propagation in the heart [11].

Na_V1.5 is located in the sarcolemma, i.e., the cardiomyocyte plasma membrane. Its localization and function are regulated by auxiliary β subunits and other associated proteins [12,13]. Alterations in the channel biophysical properties —including changes in activation or inactivation kinetics—can give rise to the channel's gain- or loss-of-function [8]. In this context, potentially fatal arrhythmias may be due to defects in regulation of transcription, affecting *SCN5A* expression, or to post-translational modifications, causing, for example, an incomplete processing of Na_V1.5 that may turn into its retention in the exocytic pathway. A precise subcellular location of Na_V1.5 is essential for the channel role in AP generation (Figure 2). Thus, defects in Na_V1.5 targeting to the sarcolemma—or an incorrect or inadequate anchoring at the membrane—can affect considerably Na_V channel functioning. In fact, alterations in these processes are associated with BrS [14].

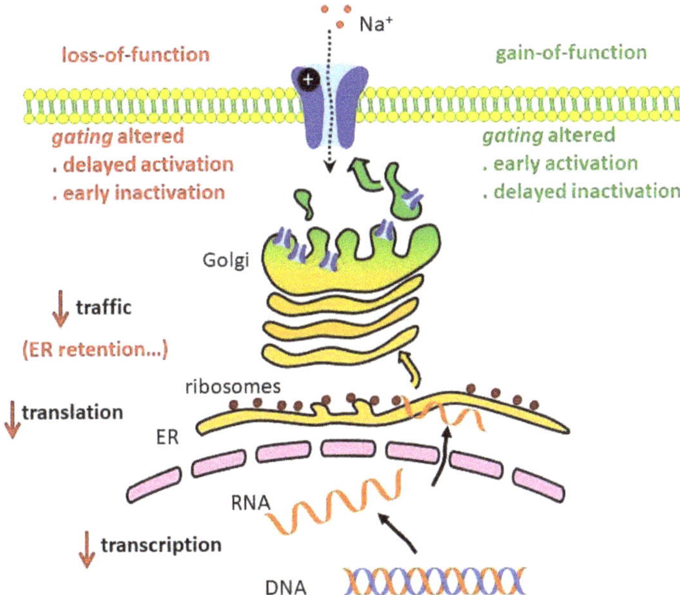

Figure 2. Alterations in the Na_V channel associated with heart disease. Several molecular mechanisms can affect expression, localization or function of the channel and are associated with inherited or acquired heart disease. ER, endoplasmic reticulum.

To understand the role of the Na_V channel in cardiac function, and the consequent alterations that may lead to disease, it is essential to study traffic and localization of Na_V1.5, and importantly, the contribution of its associated and regulatory proteins. In fact, considerable effort has been made to investigate the pathways that determine the correct location of Na_V1.5 to subregions of the sarcolemma. It is known that several proteins interact with Na_V1.5, including cytoskeleton components, and various structural domains involved in these interactions have been identified [13,15,16]. In addition, it has been reported the assembly of macromolecular complexes between Na_V1.5 and the inward rectifier potassium channel Kir2.1 (which contributes to the final repolarization phase of the AP, and controls the resting membrane potential of ventricular cardiomyocytes), which localize together in microdomains of the sarcolemma, thereby controlling excitability [17]. Interestingly, an unconventional anterograde route between Na_V1.5 and Kir2.1/2.2 has been described, which would allow reciprocal regulation of their localization. Understanding these interactions should help determining defects in localization of these channels at the sarcolemma. This is important to subsequently assess the clinical features and severity of the disease. For instance, depending on the BrS-associated mutation in *SCN5A*, a different pathological retention of Na$_V$1.5 may be observed, either in the endoplasmic reticulum (ER) or in the Golgi apparatus. In this regard, ER-, but not Golgi-retained Na$_V$1.5 mutants, can be partially restored by Kir2.1/2.2, thereby ameliorating the I$_{Na}$ reduction due to the *SCN5A* mutation. Thus, Golgi-retained Na$_V$1.5 mutants give a comparatively higher decrease in I$_{Na}$, and therefore of cardiac excitability, than the ER-retained mutants [14].

Na$_V$1.5 is located at intercalated discs (ID), that is, the region between cardiomyocytes containing cell adhesion complexes, and also in the lateral membrane [13], where costameres and transverse tubules (T tubules) are found; costameres are protein complexes connecting the cardiomyocyte sarcomeres with the extracellular matrix, while T tubules are deep invaginations of the sarcolemma in the Z-disk region (or Z-line, which borders the sarcomeres), allowing electrical coupling with Ca^{2+} release from the sarcoplasmic reticulum [18]. The distribution of Na$_V$1.5 in ID and T tubules depends in part on ankyrin-G (AnkG) [19]. In fact, multiple interactions implicating Na$_V$1.5 have been described, which establish protein complexes defining different pools of Na$_V$1.5 in cardiomyocytes (Figure 3). Thus, plakophilin 2 (PKP2) and SAP97 (synapse-associated protein 97; a member of the family of membrane-associated guanylate kinases) would control Na$_V$1.5 localization at the ID [20]. It has been proposed that PKP2 is part of a complex with connexin 43 (Cx43) and AnkG, possibly independent of the interaction with SAP97 [21]. Na$_V$1.5 is also located in the lateral membrane, whose targeting is regulated by the syntrophin/dystrophin complex connected to the actin cytoskeleton [20]. AnkG is also associated with Na$_V$1.5 in T tubules [22]. Yet, AnkG is clearly required for Na$_V$1.5 targeting to the ID [23]. Finally, it was shown by super-resolution microscopy that a complex formed by Cx43, PKP2 and N-cadherin is responsible for the delivery of microtubule cargo to the ID, including Na$_V$1.5; the components and location of this molecular complex define the so-called connexome, which regulates electrical coupling, cell adhesion, and cell excitability [24].

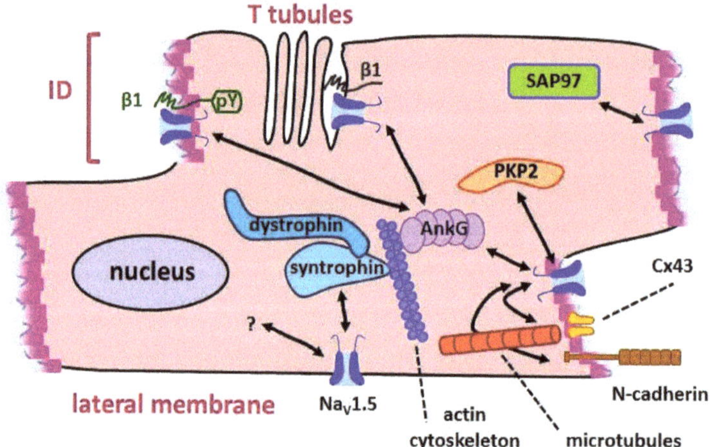

Figure 3. Multiple interactions implicating Na$_V$1.5 in cardiomyocytes. Several proteins have been identified interacting with Na$_V$1.5 in different subregions of the cardiomyocyte sarcolemma, establishing protein complexes that define different pools of Na$_V$1.5. See the text for details.

Understanding the distribution of Na$_V$1.5 in sarcolemma subdomains has indeed become a complex task. Thus, recent data suggest the presence of a sub-pool of Na$_V$1.5 at the lateral membrane that is independent of syntrophin [25]. Interestingly, even the associated β1 subunit (see below) appears differentially localized at sarcolemma subdomains. In this regard, Tyr-phosphorylated β1 (pYβ1) is associated with Na$_V$1.5 at the ID, along with other partners, such as AnkG, Cx43 and N-cadherin, whereas non-phosphorylated β1 associates with Na$_V$ channels in T tubules [26].

In summary, Na$_V$1.5 is probably grouped into well-defined functional nanodomains in each subregion [27]. In fact, Na$_V$1.5 has been found associated with lipid rafts, which are membrane domains rich in cholesterol and glycosphingolipids where ionic channel regulatory proteins concentrate [18]. Understanding how all these molecules interact to regulate the subcellular localization of Na$_V$ channels is clearly a challenge in this field of research.

1.3. The Na$_V$ Channel β Subunits

Analysis of Na$_V$1.5 trafficking can be envisaged from at least three standpoints. First, we must understand how Na$_V$1.5 is targeted to cell surface domains; secondly, how Na$_V$1.5 is retained in certain subregions or domains of the sarcolemma, thereby defining its surface distribution; and third, how Na$_V$1.5 endocytosis and turnover are regulated, allowing renewal of the channel at the surface. In this review, we will deal primarily with the first two points, focusing on a Na$_V$ channel-associated β subunit, namely β2. The β subunit family consists of four genes, *SCN1B-SCN4B*, which encode four different proteins, β1–4, of which β1 has two alternative splice variants, β1A and β1B; β subunits regulate sodium current (I$_{Na}$) density [28]. Interacting with Na$_V$1.5 by its extracellular region [29], or even through the transmembrane domain (TMD) [30], it has been proposed that they perform a major role mainly in ensuring efficient transport of the channel's α subunit to the plasma membrane [13]. Interestingly, several BrS-associated mutations causing loss-of-function of the Na$_V$ channel have been found in β subunits [31–34].

The current view on the function of β subunits within the Na$_V$ channel is that (i) they determine proper cell surface localization of the channel complex, (ii) they control channel gating and kinetics, and also (iii) they regulate gene expression of the α subunit, and perhaps of the β subunits themselves [12]. The mechanisms by which β subunits may carry into effect these roles have been addressed in excellent recent reviews [35–37]. However, the data lead to the conclusion that modulation of Na$_V$ channels by β subunits in heterologous systems shows effects that are dependent on the β and α subunits analyzed and, importantly, also of the cell type, which may be related to endogenous expression of some of the subunits. Indeed, it has been seen that in neurons or cardiomyocytes obtained from null mice, which would be more physiologically relevant models, the reported effects are in general less striking. Moreover, the protein domain(s) of β interacting with α, and posttranslational modifications, such as glycosylation and phosphorylation, clearly have an effect. In this regard, glycosylation of β subunits, and in particular sialylation, was found important for regulating the channel biophysical properties. Specifically, only sialylated β1 shifted gating of Na$_V$1.2 and Na$_V$1.5. On the other hand, sialylation of β2 influenced Na$_V$1.5 gating, while not that of Na$_V$1.2; those data also implicated that effects by these subunits on α are synergistic and, for β2 in particular, isoform specific [38].

The possibility that β subunits may compensate for each other in vivo adds another level of complexity to this aspect. In addition, Na$_V$ β subunits can also influence certain potassium currents, thereby contributing to cross-talk between Na$_V$ and potassium channels. Progress on this topic may be accomplished by investigating the structural determinants of the interaction and affinity between α and β subunits within the Na$_V$ channel [37,39]. For instance, non-covalently bound β1 and β3 subunits affect the conformational dynamics of Na$_V$1.5 by binding, through their extracellular and TMDs, to the voltage-sensing domains within domains III and IV of Na$_V$1.5, thereby modulating ionic current kinetics and cell excitability [30].

Next, we will focus specifically on regulation of β2 trafficking, and how that could influence localization of Na$_V$1.5, an aspect of key importance in cellular excitability and electrical coupling of the heart.

2. The β2 Subunit

2.1. Sequence and Domain Architecture of β2

Proper traffic and localization of β2 to subdomains of the plasma membrane is likely determined by its sequence motifs. Moreover, the interaction between β2 and Na$_V$1.5, as well as with other channel-regulatory molecules, must be relevant to determine the correct localization of Na$_V$1.5, and therefore the functionality of the Na$_V$ channel. To examine this aspect, a biochemical dissection of β2 should be performed. β2 is a type I transmembrane protein with an extracellular, immunoglobulin (Ig)-like loop, having a role in cell adhesion [12], a single TMD, and a short cytoplasmic tail [40] (Figure 4). The extracellular loop, maintained by an intramolecular disulfide bond between Cys-50 and Cys-127 [29], has three potential *N*-glycosylation sites, that is, Asn-42, Asn-66 and Asn-74 [38]. Within this region, a third cysteine, Cys-55, establishes a disulfide bond with the α subunit [29]. On the other hand, the short intracellular C-terminal domain has two possible phosphorylation sites, i.e., Ser-192 and Thr-204 [41] (based on UniProtKB accession number O60939, corresponding to human *SCN2B*).

2.2. Trafficking of β2 and Its Role within the Na$_V$ Channel

The case of β2 is of particular interest concerning control of channel localization, since it is believed to influence localization of Na$_V$1.5 at a post-Golgi site, on its targeting to the cell surface [42,43]. In addition, we have described the first mutation associated with BrS in *SCN2B* [32], the gene encoding β2. Our data have shown that the β2 D211G mutation (a substitution of Asp by Gly) causes an I$_{Na}$ reduction of ~ 40% due to decreased cell surface levels of Na$_V$1.5 [32,44]. Consistent with these data, it has been found that *Scn2b* deletion in mice entails, both in ventricular myocytes [45] and in primary cultures of hippocampal neurons [46], a comparable 40% decrease in surface levels of the α subunit and, as a result, of the I$_{Na}$. Notably, β2 must associate with the α subunit for targeting to the nodes of Ranvier and the axon initial segment of the α/β2 complex [29].

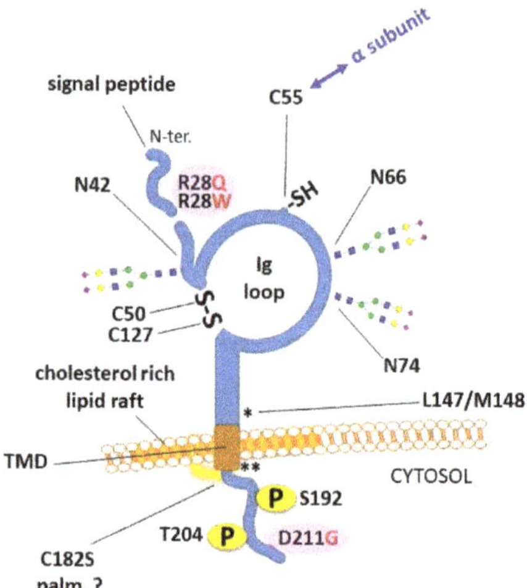

Figure 4. Sequence and domain architecture of the β2 subunit. Of its 215 amino acids, the first 29 correspond to the predicted N-terminal (N-ter.) signal peptide [41]. It is worth noting the extracellular immunoglobulin (Ig)-like loop. The transmembrane domain (TMD), which expands along ~ 20 residues, and probably associates with lipid rafts, is followed by a short C-terminal domain. Asterisks denote the approximate location of cleavage sites by β-secretase (*) and γ-secretase (**). Pathogenic mutations, i.e., the atrial fibrillation-associated R28Q and R28W, and the BrS-associated D211G, are shown in pink background; see the text for more details.

We have shown that β2, exogenously expressed in Madin-Darby canine kidney (MDCK) cells, localizes in a polarized way at the apical plasma membrane domain [44]. In both MDCK cells and in cardiomyocyte-derived HL-1 cells, localization of $Na_V1.5$ at the surface is strengthened by β2, but not by the subunit carrying the D211G BrS mutation. Regarding differential surface distribution, it is noteworthy the case of the β1 subunit, whose Tyr phosphorylation regulates its localization in sarcolemmal subdomains. As shown in mouse ventricular myocytes, Na_V channels at the ID consist of $Na_V1.5$ and pYβ1, in close association with N-cadherin and Cx43, while channels in T tubules would be formed by other Na_V isoforms, such as $Na_V1.1$ and $Na_V1.6$, linked with unphosphorylated β1, and associated with ankyrin-B [26]. Understanding the mechanisms that determine these interactions is of the utmost importance, since the differential location of Na_V channels in subregions of the sarcolemma certainly influences conduction velocity and cardiac impulse propagation [20] (see Figure 3).

It is not known how β2 is transported to the cell surface; more specifically, it remains to be addressed how β2 is targeted preferentially to the apical region in MDCK cells. In fact, understanding how apical targeting signals are recognized in proteins is the focus of intense study. Such recognition can take place by association of the protein's TMD with lipid rafts. It can also occur via *N*- or *O*-glycosylation of the luminal domain, and its consequent interaction with galectins (proteins that bind glycoprotein carbohydrates). Additional elements have been implicated, including certain Rab GTPases (belonging to the Ras superfamily of small GTPases), microtubule motor proteins, and elements of the actin cytoskeleton [47,48].

Based on these considerations, it is tempting to speculate that apical β2 localization may have a parallelism with its location at the nodes of Ranvier and the axon initial segment, where it interacts with the α subunit [29]. If β2 similarly distributes in specialized sarcolemmal regions, it would help to define, at least in part, the precise $Na_V1.5$ location also at the cardiomyocyte surface. It is important to note that it remains to be investigated whether targeting of β2 to such surface regions, and its surface dynamics, potentially involving its endocytosis and turnover, are regulated by remodeling of the lipid bilayer, glycosylation, and/or dimerization/oligomerization. In this regard, we have recently found that *N*-glycosylation of β2 is indeed necessary for its efficient trafficking and localization to the plasma membrane; importantly, non-glycosylated β2 does not promote surface localization of $Na_V1.5$ [49].

2.3. Functions of β2

As demonstrated in *Scn2b* knockout mice, β2 has essential in vivo functions in maintenance of neuronal excitability. However, it was not found necessary for Na_V channel expression, or for survival, since knockouts develop normally, their neuronal function and morphology appear normal, and also their life expectancy. Yet, β2 is necessary for voltage-dependent inactivation of the I_{Na} and for maintenance of Na_V channel levels at the plasma membrane of the neuronal soma and in myelinated axons. The lack of β2 leads to hyperexcitability and to a decrease of the threshold to suffer seizures, possibly due to reduced excitability of inhibitory interneurons [46].

Concerning cardiac function, *Scn2b* knockout mice suffer from ventricular and atrial arrhythmias [45], which is consistent with *SCN2B* mutations described in humans [32,50]. In fact, these mice seem to reproduce some aspects of BrS, and it has been proposed that they could be very useful for modeling aspects of human ventricular arrhythmias with a greater susceptibility to atrial arrhythmia [45]. Intriguingly, a more recent report conclusively demonstrates that both β2 and β4 (unlike β1 and β3) are virtually undetected in ventricle of canine heart, implying that these subunits unlikely have any contribution to the regional manifestation of BrS, typically affecting the right ventricle rather than the left one [51].

Interestingly, human β2 can be sequentially cleaved by secretases; cleavage of β2 takes place analogously to the processing of the amyloid precursor protein (APP), whose amyloidogenic cleavage is cause and aggravation of the pathogenesis in Alzheimer's disease [52]. In fact, β2 cleavage by α-secretase, and subsequently by γ-secretase, appears required for cell-cell adhesion and migration of β2-expressing cells [53]. In this regard, its extracellular domain, which confers β2 features of cell adhesion molecule (CAM), has been shown to interact via homophilic, and also heterophilic connections, by which it would bind the extracellular matrix proteins tenascin-R and tenascin-C, and also the β1 subunit, thereby functioning as a CAM [54,55]. Because β subunits are cell surface glycoproteins expressed during key periods of neuronal development, those earlier studies led to the hypothesis that they likely have a dual role, i.e., regulating the ion channel and functioning as CAMs. Such role on cell adhesion appears independent of their action on the channel α subunit, and would be responsible of bringing out an important connection between the extracellular environment and the cytoskeleton, for instance, by means of AnkG recruitment, which could have important implications in neural development [56].

The γ-secretase cleavage site in β2 [57], and probably that for β-secretase as well, appears to be conserved [58,59]. At least in mouse, the four β subunits (β1-4) can actually be processed by β-secretase, i.e., the β-site APP cleaving enzyme (BACE1) [59]. The cleaved intracellular domain of β2, as shown in neurons, is translocated to the nucleus to participate in transcriptional regulation of the α subunit, and possibly of other genes [52]. These results have been linked with observations from the null phenotype in mice. Since the *Scn2b* knockout mice displays uncontrolled fibrosis, i.e., cell replacement with fibrous connective tissue, it has been suggested that the β2 intracellular domain may regulate genes inhibiting atrial fibrosis [45].

Such remarkable finding reported over 10 years ago led to the conclusion that the β2 intracellular domain acts as a transcriptional activator for $Na_V1.1$, although the increased α subunit accumulated intracellularly, while its cell surface levels were reduced [52]. Subsequently, these authors reported that BACE1 null mice have increased $Na_V1.2$ surface levels, explained as a potential compensatory mechanism for the reduced surface $Na_V1.1$ level. However, no changes in mRNA levels for $Na_V1.2$ were detected [60]. This issue remains controversial as the phenotype of BACE1 null mice, i.e., susceptibility to seizures, could not be explained by alteration in expression and localization of $Na_V1.2$ or $Na_V1.6$, since protein levels of these α subunits remained unaltered and their distribution could not be correlated with the phenotype of these mice [61]. Interestingly, further work has shown that BACE1 can also have a non-enzymatic role on α, implying an influence on the electrical behavior of excitable cells independent of its proteolytic activity. Specifically, BACE1 was found to act as an accessory subunit of $Na_V1.2$, mimicking β2 in the α/β channel complex. This feature of BACE1, i.e., acting as an accessory β subunit of Na_V, and also potassium channels, would influence channel function, having an impact on cellular excitability and brain network activity, and at the same time may help explaining previous data. Importantly, this aspect opens a new area of study of BACE1 in physiology and pathology (reviewed in [62]).

Altogether, the published data provide strong evidence supporting a role of β subunits as signaling molecules involved in cell adhesion, migration, and neurite outgrowth [12]. In this regard, β1 was conclusively shown to promote neurite outgrowth in vivo through a signaling process triggered by trans-homophilic cell adhesion and association with lipid rafts, which was suggested to be essential for postnatal development of the central nervous system [63]. Within the context of cardiac function, an important result in this regard is the recent finding showing that β1, by mediating cell-cell adhesion, contributes to spreading of the AP along ventricular cardiomyocytes [64]. This work shows that β1 facilitates coupling of adjacent cells for ion exchange, being located within clefts of the perinexus, i.e., the narrow stretch of membranes closely apposed and adjacent to the gap junctions. By super-resolution optical microscopy, the authors concluded that β1 is located within $Na_V1.5$-enriched nanodomains of the ID, however, it was seen clearly away from the fascia adherens areas rich in N-cadherin [64].

Here, we propose that β2 may have similar, yet to be discovered functions in adhesion and signaling also in cardiomyocytes. Investigating these potential features could help to understand better the role of β2 in specialized subdomains of the sarcolemma, both within and outside the Na_V channel. Indeed, β2 also binds laminin, which is the most abundant CAM in the extracellular matrix of the peripheral nervous system. In this regard, it has been suggested that β2 overexpression in prostate cancer mediates the neoplastic invasion of the nerves, which is believed to be a common pathway for cancer metastasis [65]. Intriguingly, this feature of β2 may have a connection with previous observation showing that β2 overexpression causes an increase of the cell surface membrane in Xenopus oocytes, which was then explained by increasing fusion of intracellular vesicles with the plasma membrane [40].

3. Overview and Working Hypotheses

In the context of the heart, the current data support the idea that β2 has a role in regulating localization and function of the Na$_V$ channel in cardiomyocytes. In MDCK cells, a model system that we have used to imitate the polarized cardiomyocytes [18], β2 is located at the apical surface [44]. This distribution may have a correspondence with its preferential targeting to the neural nodes of Ranvier and the axon initial segment, where it interacts with the α subunit [29]. Given the importance of the α/β2 link, it is fundamental to understand how alterations in β2 may affect its localization and that of Na$_V$1.5 at the cell surface. In this regard, we have analyzed the influence of β2 on localization and function of the Na$_V$ channel in polarized MDCK cells. Our data show that targeting of β2 to the cell surface is affected if *N*-glycosylation is abolished, causing retention of β2 in the ER. A small fraction of the triple non-glycosylated β2 mutant, that is, N42Q/N66Q/N74Q, can still reach the cell surface by bypassing the Golgi complex, and is correctly targeted to the apical domain. However, it is important to underline that it does not promote surface localization of Na$_V$1.5, since coexpression of this mutant causes a reduction in surface levels of Na$_V$1.5 [49] (Figure 5). Therefore, it is clearly defective in regulating Na$_V$1.5 surface localization, which is a well-accepted function of β2 in the heart [12]. Yet, we have seen that a single *N*-glycosylation is sufficient for β2 to reach the cell surface and to ensure efficient Na$_V$1.5 localization [49]. To perform this role, we suggest that *N*-glycosylation in β2 is necessary for its recognition through galectins, which would guide β2 to the proper cell surface domain [66].

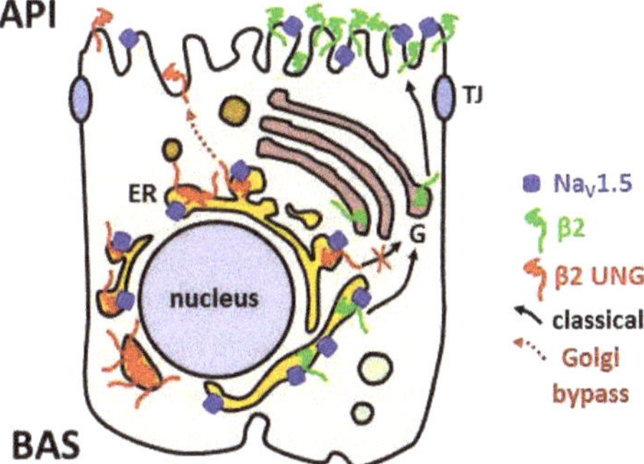

Figure 5. *N*-glycosylation of β2 is necessary for its targeting to the cell surface and to promote surface localization of Na$_V$1.5. Proper β2 trafficking is determinant for Na$_V$1.5 localization to the cell surface, and specifically, to the apical plasma membrane (API) in polarized MDCK cells. Unglycosylated (UNG; N42Q/N66Q/N74Q) β2 is retained in the endoplasmic reticulum (ER), often seen together with Na$_V$1.5. By bypassing the Golgi apparatus (G), a fraction of the mutant can reach the cell surface (dotted arrow), but at a rate of only approximately one-third of that of the wild type. In addition, it is defective in promoting surface localization of Na$_V$1.5. BAS, basolateral surface; TJ, tight junctions.

For the β1 subunit, it has been proposed that palmitoylation is a possible mechanism for its adequate surface localization and consequent anchorage to lipid rafts [28]. Along these lines, we suggest that the TMD influences the localization of β2 within surface subdomains, which would also take place by association with these cholesterol-rich regions, and would possibly be regulated by the TMD length [67]. In this regard, our ongoing work shows that the apical distribution of β2 in MDCK cells is altered upon cholesterol depletion [68]. In primary cortical neurons, all four β subunits (β1-4) have in fact been found enriched in detergent resistant membranes by subcellular fractionation, implying their preferential association with lipid rafts [59]. On the other hand, it remains to be addressed the possible contribution of the Ig domain [29,69] in disulfide bond-mediated dimerization/oligomerization of β2 [70] and, consequently, in β2 targeting to certain surface domains.

We propose that mutations associated with β2—some of which undoubtedly linked to disease [32,50]—alter its dynamics at the membrane, negatively affecting localization and function of the Na$_V$ channel. As previously seen [44–46], mutated β2, or its absence, negatively affects Na$_V$1.5 traffic from the ER to the plasma membrane. A mutation in β2 may cause its intracellular retention. In the case of the BrS-associated D211G mutation, β2 reaches the plasma membrane normally. However, similar to unglycosylated β2 [49], this mutant is defective in promoting Na$_V$1.5 trafficking to the surface [44]. In addition to β2 *N-glycosylation*, which should ensure Na$_V$1.5 chaperoning for proper folding, we propose that there are elements in the cytoplasmic domain of β2 potentially implicated on regulating Na$_V$1.5 localization. In this regard, dynamics in β2 phosphorylation could potentially influence recruitment of adapter proteins, such as AnkG [19], thereby affecting targeting of the Na$_V$ channel to the right surface subdomain. Given its proximity, the D211G mutation could influence potential phosphorylation predicted to take place in C-terminal residues [41] (see Figure 4). Support for this idea comes from previous work showing that the β2 intracellular domain, possibly interacting with AnkG, allows the heterophilic interaction with the extracellular domain of β1, thereby stabilizing extracellular cell adhesion. The region in β2 required was mapped, and encompasses two Thr residues, i.e., T198 and T204, that belong to a consensus sequence for potential phosphorylation by casein kinase II, and thereby may be critical for AnkG interaction, in analogy to Tyr-phosphorylated β1 [71,72].

Additional evidence suggesting that there is an indirect link between Na$_V$1.5 and β subunits implicated on regulating localization of α also comes from earlier work showing that the cytoplasmic domain of β subunits may not have much influence on α/β interaction in heterologous cell systems. Thus, a β1 chimera bearing the intracellular domain of β2 overlapped strongly with Na$_V$1.5, supposedly in intracellular compartments [73], similarly as β1 does, but in contrast to β2, which is efficiently targeted to the cell surface [42]. Moreover, recent data obtained by cryo-electron microscopy confirmed that direct α/β2 interaction takes place essentially through extracellular disulfide binding, i.e., at Cys-55, as previously shown [29], in addition to hydrogen bonding with Tyr-56 and Arg-135, although the latter, connecting with the Ig loop, are likely less stable interactions [74,75]. Interestingly, a previous report presenting 3D structural data of the β2 extracellular region obtained by X-ray crystallography revealed the exact points of contact with the α subunit and indicated that β2 is unique among the Na$_V$ β subunits in that it contains a second intrasubunit disulfide bond, via Cys-72 and Cys-75, in addition to the common Cys-50 and Cys-127 bond, conserved in its β counterparts, yet this peculiarity may have a yet to be elucidated function [76]. On the whole, we suggest that, at least in part, the action of β2 on α, and in particular on Na$_V$1.5, may take place indirectly.

4. Limitations: The Cell Model System

Carrying out studies in immortalized cell lines, such as Human Embryonic Kidney 293 cells, Chinese Hamster Ovary cells, or MDCK cells, represents an obvious limitation to recapitulate the biology of the β2 subunit in the cardiomyocyte, or in other excitable cells. Like cardiomyocytes, MDCK cells are polarized, but unlike those, they are epithelial cells. On the other hand, HL-1 cells are of cardiac origin, albeit they are not polarized. In general, mechanisms governing trafficking have been studied in cell lines growing in artificial environments, such as monolayers attached on a substrate, or embedded in a matrix resembling the extracellular milieu. These setups are suitable for analyzing many aspects addressed here. However, most of these cell lines have the obvious limitation that they are not excitable. Even HL-1 cells, which have endogenous I$_{Na}$ [77], do not express β subunits and have very low levels of Na$_V$1.5 ([44], and our unpublished data), and must also be transfected to study sorting and targeting of the Na$_V$ channel. Likewise, most of these cell lines probably lack specific channel-associated proteins, as well as components of the machinery required for sorting, targeting, or membrane anchoring of channel subunits and, therefore, for their proper localization in the cardiomyocyte. Therefore, we must keep in mind that when studying the role of β subunits, their trafficking, or interaction with α, it is quite possible that some observed effects differ from what really occurs in a cardiomyocyte or in other excitable cells. Definitely, this issue should be addressed by researchers in the field, and a considerable effort must be put on developing more suitable models to analyze trafficking of the Na_V channel subunits and also of the channel's biophysical properties.

5. Implications of Research on the Na_V Channel β Subunits

Research projects strictly focused on the biological aspects, as those with a strong relevance to clinical aspects, are valid to address the aspects discussed here. Studying the mechanisms that may be altered in heart disease poses to address one of the main challenges of society, that is, human health. A key hypothesis is that alterations in localization of components of the Na_V channel complex convey a risk of arrhythmia and, therefore, are potentially associated with important cardiac pathologies.

Altogether, it is of utmost importance that findings in the field lead to a better understanding of how subunits of the Na_V channel correctly localize. Undoubtedly, growing knowledge on this aspect will provide a better picture on the link between cell excitability and the electrical coupling in the heart, thereby contributing to a better knowledge of how arrhythmias develop.

In summary, future studies should be aimed to understanding better the genetic factors that determine phenotype variability. We are convinced that data generated in this field will be very useful for risk stratification of the patient and therefore to predict heart disease. This should contribute to early diagnosis of disease in asymptomatic individuals potentially at risk of SCD and to improve treatment therapies, also avoiding a fatal event in patients already suffering from heart disease.

Author Contributions: Conceptualization, M.V. and E.C.; investigation, M.V. and E.C.; resources, R.B.; writing—original draft preparation, M.V.; writing—review and editing, M.V., E.C. and R.B.; funding acquisition, R.B.

Funding: This research and the APC were funded by "La Caixa" Foundation, a grant to R.B (number N/A).

Acknowledgments: E.C. is recipient of a predoctoral fellowship (FI_B 00071) from the Agència de Gestió d'Ajuts Universitaris i de Recerca (AGAUR)—Generalitat de Catalunya. We also thank the University of Girona and the Spanish Instituto de Salud Carlos III (ISCIII); the CIBERCV is an initiative of the ISCIII from the Spanish Ministerio de Ciencia, Innovación y Universidades.

Conflicts of Interest: The authors declare no conflict of interest. The funders had no role in the design of the study; in the collection, analyses, or interpretation of data; in the writing of the manuscript, or in the decision to publish the results.

References

1. Rubart, M.; Zipes, D.P. Mechanisms of sudden cardiac death. *J. Clin. Investig.* **2005**, *115*, 2305–2315. [CrossRef] [PubMed]
2. Mehra, R. Global public health problem of sudden cardiac death. *J. Electrocardiol.* **2007**, *40*, S118–S122. [CrossRef] [PubMed]
3. Campuzano, O.; Beltran-Alvarez, P.; Iglesias, A.; Scornik, F.; Perez, G.; Brugada, R. Genetics and cardiac channelopathies. *Genet. Med.* **2010**, *12*, 260–267. [CrossRef] [PubMed]
4. Napolitano, C.; Priori, S.G. Brugada syndrome. *Orphanet J. Rare Dis.* **2006**, *1*, 35. [CrossRef]
5. Bezzina, C.R.; Rook, M.B.; Wilde, A.A. Cardiac sodium channel and inherited arrhythmia syndromes. *Cardiovasc. Res.* **2001**, *49*, 257–271. [CrossRef]
6. Brugada, P.; Brugada, J. Right bundle branch block, persistent ST segment elevation and sudden cardiac death: A distinct clinical and electrocardiographic syndrome. A multicenter report. *J. Am. Coll. Cardiol.* **1992**, *20*, 1391–1396. [CrossRef]
7. Bastiaenen, R.; Behr, E.R. Sudden death and ion channel disease: Pathophysiology and implications for management. *Heart* **2011**, *97*, 1365–1372. [CrossRef]
8. Amin, A.S.; Asghari-Roodsari, A.; Tan, H.L. Cardiac sodium channelopathies. *Pflug. Arch.* **2010**, *460*, 223–237. [CrossRef]

9.	Wang, Q.; Shen, J.; Splawski, I.; Atkinson, D.; Li, Z.; Robinson, J.L.; Moss, A.J.; Towbin, J.A.; Keating, M.T. SCN5A mutations associated with an inherited cardiac arrhythmia, long QT syndrome. *Cell* **1995**, *80*, 805–811. [CrossRef]

10.	Ruan, Y.; Liu, N.; Priori, S.G. Sodium channel mutations and arrhythmias. *Nat. Rev. Cardiol.* **2009**, *6*, 337–348. [CrossRef]

11.	Abriel, H.; Kass, R.S. Regulation of the voltage-gated cardiac sodium channel Nav1.5 by interacting proteins. *Trends Cardiovasc. Med.* **2005**, *15*, 35–40. [CrossRef] [PubMed]

12.	O'Malley, H.A.; Isom, L.L. Sodium channel beta subunits: emerging targets in channelopathies. *Annu. Rev. Physiol.* **2015**, *77*, 481–504. [CrossRef] [PubMed]

13.	Rook, M.B.; Evers, M.M.; Vos, M.A.; Bierhuizen, M.F. Biology of cardiac sodium channel Nav1.5 expression. *Cardiovasc. Res.* **2012**, *93*, 12–23. [CrossRef]

14.	Perez-Hernandez, M.; Matamoros, M.; Alfayate, S.; Nieto-Marin, P.; Utrilla, R.G.; Tinaquero, D.; de Andres, R.; Crespo, T.; Ponce-Balbuena, D.; Willis, B.C.; et al. Brugada syndrome trafficking-defective Nav1.5 channels can trap cardiac Kir2.1/2.2 channels. *JCI Insight* **2018**, *3*, 96291. [CrossRef] [PubMed]

15.	Abriel, H. Cardiac sodium channel Na(v)1.5 and interacting proteins: Physiology and pathophysiology. *J. Mol. Cell Cardiol.* **2010**, *48*, 2–11. [CrossRef]

16.	Liu, M.; Yang, K.C.; Dudley, S.C., Jr. Cardiac sodium channel mutations: Why so many phenotypes? *Nat. Rev. Cardiol.* **2014**, *11*, 607–615. [CrossRef] [PubMed]

17.	Ponce-Balbuena, D.; Guerrero-Serna, G.; Valdivia, C.R.; Caballero, R.; Diez-Guerra, F.J.; Jimenez-Vazquez, E.N.; Ramirez, R.J.; Monteiro da Rocha, A.; Herron, T.J.; Campbell, K.F.; et al. Cardiac Kir2.1 and NaV1.5 channels traffic together to the sarcolemma to control excitability. *Circ. Res.* **2018**, *122*, 1501–1516. [CrossRef]

18.	Balse, E.; Steele, D.F.; Abriel, H.; Coulombe, A.; Fedida, D.; Hatem, S.N. Dynamic of ion channel expression at the plasma membrane of cardiomyocytes. *Physiol. Rev.* **2012**, *92*, 1317–1358. [CrossRef]

19.	Mohler, P.J.; Rivolta, I.; Napolitano, C.; LeMaillet, G.; Lambert, S.; Priori, S.G.; Bennett, V. Nav1.5 E1053K mutation causing Brugada syndrome blocks binding to ankyrin-G and expression of Nav1.5 on the surface of cardiomyocytes. *Proc. Natl. Acad. Sci. USA* **2004**, *101*, 17533–17538. [CrossRef]

20.	Petitprez, S.; Zmoos, A.F.; Ogrodnik, J.; Balse, E.; Raad, N.; El-Haou, S.; Albesa, M.; Bittihn, P.; Luther, S.; Lehnart, S.E.; et al. SAP97 and dystrophin macromolecular complexes determine two pools of cardiac sodium channels Nav1.5 in cardiomyocytes. *Circ. Res.* **2011**, *108*, 294–304. [CrossRef]

21.	Sato, P.Y.; Coombs, W.; Lin, X.; Nekrasova, O.; Green, K.J.; Isom, L.L.; Taffet, S.M.; Delmar, M. Interactions between ankyrin-G, plakophilin-2, and Connexin43 at the cardiac intercalated disc. *Circ. Res.* **2011**, *109*, 193–201. [CrossRef] [PubMed]

22.	Bennett, V.; Healy, J. Being there: Cellular targeting of voltage-gated sodium channels in the heart. *J. Cell Biol.* **2008**, *180*, 13–15. [CrossRef] [PubMed]

23.	Lowe, J.S.; Palygin, O.; Bhasin, N.; Hund, T.J.; Boyden, P.A.; Shibata, E.; Anderson, M.E.; Mohler, P.J. Voltage-gated Nav channel targeting in the heart requires an ankyrin-G dependent cellular pathway. *J. Cell Biol.* **2008**, *180*, 173–186. [CrossRef] [PubMed]

24.	Agullo-Pascual, E.; Lin, X.; Leo-Macias, A.; Zhang, M.; Liang, F.X.; Li, Z.; Pfenniger, A.; Lubkemeier, I.; Keegan, S.; Fenyo, D.; et al. Super-resolution imaging reveals that loss of the C-terminus of connexin43 limits microtubule plus-end capture and Nav1.5 localization at the intercalated disc. *Cardiovasc. Res.* **2014**, *104*, 371–381. [CrossRef] [PubMed]

25.	Rougier, J.S.; Essers, M.C.; Gillet, L.; Guichard, S.; Sonntag, S.; Shmerling, D.; Abriel, H. A distinct pool of Nav1.5 channels at the lateral membrane of murine ventricular cardiomyocytes. *Front. Physiol.* **2019**, *10*, 834. [CrossRef] [PubMed]

26.	Malhotra, J.D.; Thyagarajan, V.; Chen, C.; Isom, L.L. Tyrosine-phosphorylated and nonphosphorylated sodium channel beta1 subunits are differentially localized in cardiac myocytes. *J. Biol. Chem.* **2004**, *279*, 40748–40754. [CrossRef] [PubMed]

27.	Gillet, L.; Shy, D.; Abriel, H. Elucidating sodium channel Nav1.5 clustering in cardiac myocytes using super-resolution techniques. *Cardiovasc. Res.* **2014**, *104*, 231–233. [CrossRef]

28.	Brackenbury, W.J.; Isom, L.L. Na channel beta subunits: Overachievers of the ion channel family. *Front. Pharmacol.* **2011**, *2*, 53. [CrossRef]

29. Chen, C.; Calhoun, J.D.; Zhang, Y.; Lopez-Santiago, L.; Zhou, N.; Davis, T.H.; Salzer, J.L.; Isom, L.L. Identification of the cysteine residue responsible for disulfide linkage of Na+ channel alpha and beta2 subunits. *J. Biol. Chem.* **2012**, *287*, 39061–39069. [CrossRef]

30. Zhu, W.; Voelker, T.L.; Varga, Z.; Schubert, A.R.; Nerbonne, J.M.; Silva, J.R. Mechanisms of noncovalent beta subunit regulation of NaV channel gating. *J. Gen. Physiol.* **2017**, *149*, 813–831. [CrossRef]

31. Hu, D.; Barajas-Martinez, H.; Burashnikov, E.; Springer, M.; Wu, Y.; Varro, A.; Pfeiffer, R.; Koopmann, T.T.; Cordeiro, J.M.; Guerchicoff, A.; et al. A mutation in the beta 3 subunit of the cardiac sodium channel associated with Brugada ECG phenotype. *Circ. Cardiovasc. Genet.* **2009**, *2*, 270–278. [CrossRef]

32. Riuro, H.; Beltran-Alvarez, P.; Tarradas, A.; Selga, E.; Campuzano, O.; Verges, M.; Pagans, S.; Iglesias, A.; Brugada, J.; Brugada, P.; et al. A missense mutation in the sodium channel beta2 subunit reveals SCN2B as a new candidate gene for Brugada syndrome. *Hum. Mutat.* **2013**, *34*, 961–966. [CrossRef] [PubMed]

33. Watanabe, H.; Koopmann, T.T.; Le Scouarnec, S.; Yang, T.; Ingram, C.R.; Schott, J.J.; Demolombe, S.; Probst, V.; Anselme, F.; Escande, D.; et al. Sodium channel beta1 subunit mutations associated with Brugada syndrome and cardiac conduction disease in humans. *J. Clin. Investig.* **2008**, *118*, 2260–2268. [CrossRef] [PubMed]

34. Ishikawa, T.; Takahashi, N.; Ohno, S.; Sakurada, H.; Nakamura, K.; On, Y.K.; Park, J.E.; Makiyama, T.; Horie, M.; Arimura, T.; et al. Novel SCN3B mutation associated with brugada syndrome affects intracellular trafficking and function of Nav1.5. *Circ. J.* **2013**, *77*, 959–967. [CrossRef] [PubMed]

35. Bouza, A.A.; Isom, L.L. Voltage-gated sodium channel beta subunits and their related diseases. In *Handbook of Experimental Pharmacology*; Springer: Berlin, Germany, 2017. [CrossRef]

36. Hull, J.M.; Isom, L.L. Voltage-gated sodium channel beta subunits: The power outside the pore in brain development and disease. *Neuropharmacology* **2018**, *132*, 43–57. [CrossRef] [PubMed]

37. Molinarolo, S.; Granata, D.; Carnevale, V.; Ahern, C.A. Mining protein evolution for insights into mechanisms of voltage-dependent sodium channel auxiliary subunits. *Handb. Exp. Pharmacol.* **2018**, *246*, 33–49. [CrossRef] [PubMed]

38. Johnson, D.; Bennett, E.S. Isoform-specific effects of the beta2 subunit on voltage-gated sodium channel gating. *J. Biol. Chem.* **2006**, *281*, 25875–25881. [CrossRef] [PubMed]

39. Namadurai, S.; Yereddi, N.R.; Cusdin, F.S.; Huang, C.L.; Chirgadze, D.Y.; Jackson, A.P. A new look at sodium channel beta subunits. *Open Biol.* **2015**, *5*, 140192. [CrossRef]

40. Isom, L.L.; Ragsdale, D.S.; De Jongh, K.S.; Westenbroek, R.E.; Reber, B.F.; Scheuer, T.; Catterall, W.A. Structure and function of the beta 2 subunit of brain sodium channels, a transmembrane glycoprotein with a CAM motif. *Cell* **1995**, *83*, 433–442. [CrossRef]

41. UniProt Consortium. UniProt: A hub for protein information. *Nucleic Acids Res.* **2015**, *43*, D204–D212. [CrossRef]

42. Zimmer, T.; Biskup, C.; Bollensdorff, C.; Benndorf, K. The beta1 subunit but not the beta2 subunit colocalizes with the human heart Na+ channel (hH1) already within the endoplasmic reticulum. *J. Membr. Biol.* **2002**, *186*, 13–21. [CrossRef] [PubMed]

43. Schmidt, J.W.; Catterall, W.A. Biosynthesis and processing of the alpha subunit of the voltage-sensitive sodium channel in rat brain neurons. *Cell* **1986**, *46*, 437–444. [CrossRef]

44. Dulsat, G.; Palomeras, S.; Cortada, E.; Riuro, H.; Brugada, R.; Verges, M. Trafficking and localisation to the plasma membrane of Nav1.5 promoted by the beta2 subunit is defective due to a beta2 mutation associated with Brugada syndrome. *Biol. Cell* **2017**, *109*, 273–291. [CrossRef] [PubMed]

45. Bao, Y.; Willis, B.C.; Frasier, C.R.; Lopez-Santiago, L.F.; Lin, X.; Ramos-Mondragon, R.; Auerbach, D.S.; Chen, C.; Wang, Z.; Anumonwo, J.; et al. Scn2b deletion in mice results in ventricular and atrial arrhythmias. *Circ. Arrhythm. Electrophysiol.* **2016**, *9*. [CrossRef]

46. Chen, C.; Bharucha, V.; Chen, Y.; Westenbroek, R.E.; Brown, A.; Malhotra, J.D.; Jones, D.; Avery, C.; Gillespie, P.J., 3rd; Kazen-Gillespie, K.A.; et al. Reduced sodium channel density, altered voltage dependence of inactivation, and increased susceptibility to seizures in mice lacking sodium channel beta 2-subunits. *Proc. Natl. Acad. Sci. USA* **2002**, *99*, 17072–17077. [CrossRef]

47. Stoops, E.H.; Caplan, M.J. Trafficking to the apical and basolateral membranes in polarized epithelial cells. *J. Am. Soc. Nephrol.* **2014**, *25*, 1375–1386. [CrossRef]

48. Weisz, O.A.; Rodriguez-Boulan, E. Apical trafficking in epithelial cells: Signals, clusters and motors. *J. Cell Sci.* **2009**, *122*, 4253–4266. [CrossRef]

49. Cortada, E.; Brugada, R.; Verges, M. N-glycosylation of the voltage-gated sodium channel beta2 subunit is required for efficient trafficking of NaV1.5/beta2 to the plasma membrane. *J. Biol. Chem.* **2019**, *294*. [CrossRef]

50. Watanabe, H.; Darbar, D.; Kaiser, D.W.; Jiramongkolchai, K.; Chopra, S.; Donahue, B.S.; Kannankeril, P.J.; Roden, D.M. Mutations in sodium channel beta1- and beta2-subunits associated with atrial fibrillation. *Circ. Arrhythm. Electrophysiol.* **2009**, *2*, 268–275. [CrossRef]

51. Calloe, K.; Aistrup, G.L.; Di Diego, J.M.; Goodrow, R.J.; Treat, J.A.; Cordeiro, J.M. Interventricular differences in sodium current and its potential role in Brugada syndrome. *Physiol. Rep.* **2018**, *6*, e13787. [CrossRef]

52. Kim, D.Y.; Carey, B.W.; Wang, H.; Ingano, L.A.; Binshtok, A.M.; Wertz, M.H.; Pettingell, W.H.; He, P.; Lee, V.M.; Woolf, C.J.; et al. BACE1 regulates voltage-gated sodium channels and neuronal activity. *Nat. Cell Biol.* **2007**, *9*, 755–764. [CrossRef] [PubMed]

53. Kim, D.Y.; Ingano, L.A.; Carey, B.W.; Pettingell, W.H.; Kovacs, D.M. Presenilin/gamma-secretase-mediated cleavage of the voltage-gated sodium channel beta2-subunit regulates cell adhesion and migration. *J. Biol. Chem.* **2005**, *280*, 23251–23261. [CrossRef] [PubMed]

54. Xiao, Z.C.; Ragsdale, D.S.; Malhotra, J.D.; Mattei, L.N.; Braun, P.E.; Schachner, M.; Isom, L.L. Tenascin-R is a functional modulator of sodium channel beta subunits. *J. Biol. Chem.* **1999**, *274*, 26511–26517. [CrossRef] [PubMed]

55. Srinivasan, J.; Schachner, M.; Catterall, W.A. Interaction of voltage-gated sodium channels with the extracellular matrix molecules tenascin-C and tenascin-R. *Proc. Natl. Acad. Sci. USA* **1998**, *95*, 15753–15757. [CrossRef]

56. Malhotra, J.D.; Kazen-Gillespie, K.; Hortsch, M.; Isom, L.L. Sodium channel beta subunits mediate homophilic cell adhesion and recruit ankyrin to points of cell-cell contact. *J. Biol. Chem.* **2000**, *275*, 11383–11388. [CrossRef]

57. Chopra, S.S.; Watanabe, H.; Zhong, T.P.; Roden, D.M. Molecular cloning and analysis of zebrafish voltage-gated sodium channel beta subunit genes: Implications for the evolution of electrical signaling in vertebrates. *BMC Evol. Biol.* **2007**, *7*, 113. [CrossRef]

58. Gersbacher, M.T.; Kim, D.Y.; Bhattacharyya, R.; Kovacs, D.M. Identification of BACE1 cleavage sites in human voltage-gated sodium channel beta 2 subunit. *Mol. Neurodegener.* **2010**, *5*, 61. [CrossRef]

59. Wong, H.K.; Sakurai, T.; Oyama, F.; Kaneko, K.; Wada, K.; Miyazaki, H.; Kurosawa, M.; De Strooper, B.; Saftig, P.; Nukina, N. beta Subunits of voltage-gated sodium channels are novel substrates of beta-site amyloid precursor protein-cleaving enzyme (BACE1) and gamma-secretase. *J. Biol. Chem.* **2005**, *280*, 23009–23017. [CrossRef]

60. Kim, D.Y.; Gersbacher, M.T.; Inquimbert, P.; Kovacs, D.M. Reduced sodium channel Na(v)1.1 levels in BACE1-null mice. *J. Biol. Chem.* **2011**, *286*, 8106–8116. [CrossRef]

61. Hitt, B.D.; Jaramillo, T.C.; Chetkovich, D.M.; Vassar, R. BACE1-/- mice exhibit seizure activity that does not correlate with sodium channel level or axonal localization. *Mol. Neurodegener.* **2010**, *5*, 31. [CrossRef]

62. Lehnert, S.; Hartmann, S.; Hessler, S.; Adelsberger, H.; Huth, T.; Alzheimer, C. Ion channel regulation by beta-secretase BACE1 - enzymatic and non-enzymatic effects beyond Alzheimer's disease. *Channels* **2016**, *10*, 365–378. [CrossRef] [PubMed]

63. Brackenbury, W.J.; Davis, T.H.; Chen, C.; Slat, E.A.; Detrow, M.J.; Dickendesher, T.L.; Ranscht, B.; Isom, L.L. Voltage-gated Na+ channel beta1 subunit-mediated neurite outgrowth requires Fyn kinase and contributes to postnatal CNS development in vivo. *J. Neurosci.* **2008**, *28*, 3246–3256. [CrossRef] [PubMed]

64. Veeraraghavan, R.; Hoeker, G.S.; Alvarez-Laviada, A.; Hoagland, D.; Wan, X.; King, D.R.; Sanchez-Alonso, J.; Chen, C.; Jourdan, J.; Isom, L.L.; et al. The adhesion function of the sodium channel beta subunit (beta1) contributes to cardiac action potential propagation. *eLife* **2018**, *7*, e37610. [CrossRef] [PubMed]

65. Jansson, K.H.; Castillo, D.G.; Morris, J.W.; Boggs, M.E.; Czymmek, K.J.; Adams, E.L.; Schramm, L.P.; Sikes, R.A. Identification of beta-2 as a key cell adhesion molecule in PCa cell neurotropic behavior: A novel ex vivo and biophysical approach. *PLoS ONE* **2014**, *9*, e98408. [CrossRef]

66. Castillon, G.A.; Michon, L.; Watanabe, R. Apical sorting of lysoGPI-anchored proteins occurs independent of association with detergent-resistant membranes but dependent on their N-glycosylation. *Mol. Biol. Cell* **2013**, *24*, 2021–2033. [CrossRef]

67. Sharpe, H.J.; Stevens, T.J.; Munro, S. A comprehensive comparison of transmembrane domains reveals organelle-specific properties. *Cell* **2010**, *142*, 158–169. [CrossRef]

68. Cortada, E.; Serradesanferm, R.; Brugada, R.; Verges, M. Role of lipid raft domains in plasma membrane dynamics of the voltage-gated sodium channel β2 subunit. (unpublished work).
69. Wimmer, V.C.; Reid, C.A.; Mitchell, S.; Richards, K.L.; Scaf, B.B.; Leaw, B.T.; Hill, E.L.; Royeck, M.; Horstmann, M.T.; Cromer, B.A.; et al. Axon initial segment dysfunction in a mouse model of genetic epilepsy with febrile seizures plus. *J. Clin. Investig.* **2010**, *120*, 2661–2671. [CrossRef]
70. Namadurai, S.; Balasuriya, D.; Rajappa, R.; Wiemhofer, M.; Stott, K.; Klingauf, J.; Edwardson, J.M.; Chirgadze, D.Y.; Jackson, A.P. Crystal structure and molecular imaging of the Nav channel beta3 subunit indicates a trimeric assembly. *J. Biol. Chem.* **2014**, *289*, 10797–10811. [CrossRef]
71. McEwen, D.P.; Isom, L.L. Heterophilic interactions of sodium channel beta1 subunits with axonal and glial cell adhesion molecules. *J. Biol. Chem.* **2004**, *279*, 52744–52752. [CrossRef]
72. McEwen, D.P.; Meadows, L.S.; Chen, C.; Thyagarajan, V.; Isom, L.L. Sodium channel beta1 subunit-mediated modulation of Nav1.2 currents and cell surface density is dependent on interactions with contactin and ankyrin. *J. Biol. Chem.* **2004**, *279*, 16044–16049. [CrossRef]
73. Zimmer, T.; Benndorf, K. The intracellular domain of the beta 2 subunit modulates the gating of cardiac Nav1.5 channels. *Biophys. J.* **2007**, *92*, 3885–3892. [CrossRef] [PubMed]
74. Shen, H.; Liu, D.; Wu, K.; Lei, J.; Yan, N. Structures of human Nav1.7 channel in complex with auxiliary subunits and animal toxins. *Science* **2019**, *363*, 1303–1308. [CrossRef]
75. Pan, X.; Li, Z.; Huang, X.; Huang, G.; Gao, S.; Shen, H.; Liu, L.; Lei, J.; Yan, N. Molecular basis for pore blockade of human Na(+) channel Nav1.2 by the mu-conotoxin KIIIA. *Science* **2019**, *363*, 1309–1313. [CrossRef] [PubMed]
76. Das, S.; Gilchrist, J.; Bosmans, F.; Van Petegem, F. Binary architecture of the Nav1.2-beta2 signaling complex. *eLife* **2016**, *5*. [CrossRef] [PubMed]
77. Claycomb, W.C.; Lanson, N.A., Jr.; Stallworth, B.S.; Egeland, D.B.; Delcarpio, J.B.; Bahinski, A.; Izzo, N.J., Jr. HL-1 cells: A cardiac muscle cell line that contracts and retains phenotypic characteristics of the adult cardiomyocyte. *Proc. Natl. Acad. Sci. USA* **1998**, *95*, 2979–2984. [CrossRef]

 *biomolecules*

Review

Cell-Adhesion Properties of β-Subunits in the Regulation of Cardiomyocyte Sodium Channels

Samantha C. Salvage [1,*], Christopher L.-H. Huang [1,2] and Antony P. Jackson [1,*]

[1] Department of Biochemistry, University of Cambridge, Cambridge CB2 1QW, UK; clh11@cam.ac.uk
[2] Department of Physiology, Development and Neuroscience, University of Cambridge, Cambridge CB2 3EG, UK
* Correspondence: ss2148@cam.ac.uk (S.C.S.); apj10@cam.ac.uk (A.P.J.); Tel.: +44-1223-765950 (S.C.S.); +44-1223-765951 (A.P.J.)

Received: 7 June 2020; Accepted: 27 June 2020; Published: 1 July 2020

Abstract: Voltage-gated sodium (Nav) channels drive the rising phase of the action potential, essential for electrical signalling in nerves and muscles. The Nav channel α-subunit contains the ion-selective pore. In the cardiomyocyte, Nav1.5 is the main Nav channel α-subunit isoform, with a smaller expression of neuronal Nav channels. Four distinct regulatory β-subunits (β1–4) bind to the Nav channel α-subunits. Previous work has emphasised the β-subunits as direct Nav channel gating modulators. However, there is now increasing appreciation of additional roles played by these subunits. In this review, we focus on β-subunits as homophilic and heterophilic cell-adhesion molecules and the implications for cardiomyocyte function. Based on recent cryogenic electron microscopy (cryo-EM) data, we suggest that the β-subunits interact with Nav1.5 in a different way from their binding to other Nav channel isoforms. We believe this feature may facilitate *trans*-cell-adhesion between β1-associated Nav1.5 subunits on the intercalated disc and promote ephaptic conduction between cardiomyocytes.

Keywords: voltage-gated sodium (Nav) channels; Nav1.5; sodium (Nav) channel β-subunits; cell-adhesion; ephaptic conduction

---

1. Introduction

Cardiomyocytes within cardiac muscle bundles perform the involuntary contraction and relaxation cycle that is the cellular basis of the heartbeat. Cardiomyocytes possess unique adaptations to ensure this process is tightly synchronised between individual cells. In particular, the cardiomyocytes are both physically and electrically connected to each other via their intercalated discs. On the lateral membrane, T-tubules facilitate the transmission of the electrical signal from the cell surface, to deeper within the cell. This stimulates the release of calcium from the sarcoplasmic reticulum and the initiation of sarcomere contraction (Figure 1) [1].

The cardiac action potential underlies electrical signalling and is initiated by the transient depolarisation of voltage-gated sodium (Nav) channels (for further details, see Ref. [2], this volume). The Nav channel α-subunit (Mwt ~220–250 kDa) contains the ion-selective pore. In the human genome, there are nine different functional Nav channel α-subunit genes encoding proteins Nav1.1-1.9. Different Nav channel α-subunit isoforms are expressed in a tissue-specific manner and exhibit distinct gating behaviour, presumably tailored to their physiological context. In the cardiomyocyte, Nav channels with different gating properties can also be correlated with their differing sub-cellular localisation. The major Nav channel isoform expressed in the heart is Nav1.5. It is mainly localised at the intercalated disc and within caveolae on the sarcolemmal lateral membrane [3]. Cardiomyocytes also express smaller amounts of the neuronal channels Nav1.1, Nav1.3 and Nav1.6, which are predominantly localised in the T-tubules [4,5]. This pattern is striking and is likely to be functionally significant.

For example, on a given cardiomyocyte, all Nav channels will experience the same resting potential. However, Nav1.5 activates at more negative potentials and more slowly compared to neuronal Nav channels. Thus, Nav1.5 at the intercalated disc and on the sarcolemma may initiate the cardiac action potential as it propagates from one cardiomyocyte to another within the muscle fibre [5,6]. By contrast, a delayed T-tubular excitation of the neuronal Nav channels will be matched by their more negative threshold for excitation and the more rapid kinetics of activation. This, combined with the close structural association between the neuronal Nav channels, the sodium-calcium exchanger (NCX) and the voltage-gated calcium channels on the T-tubular membrane and with the ryanodine receptors (RyR) on the adjacent sarcoplasmic reticulum, permits T-tubular activation that is synchronous with the surface action potential and that optimally initiates excitation-contraction coupling [4,7].

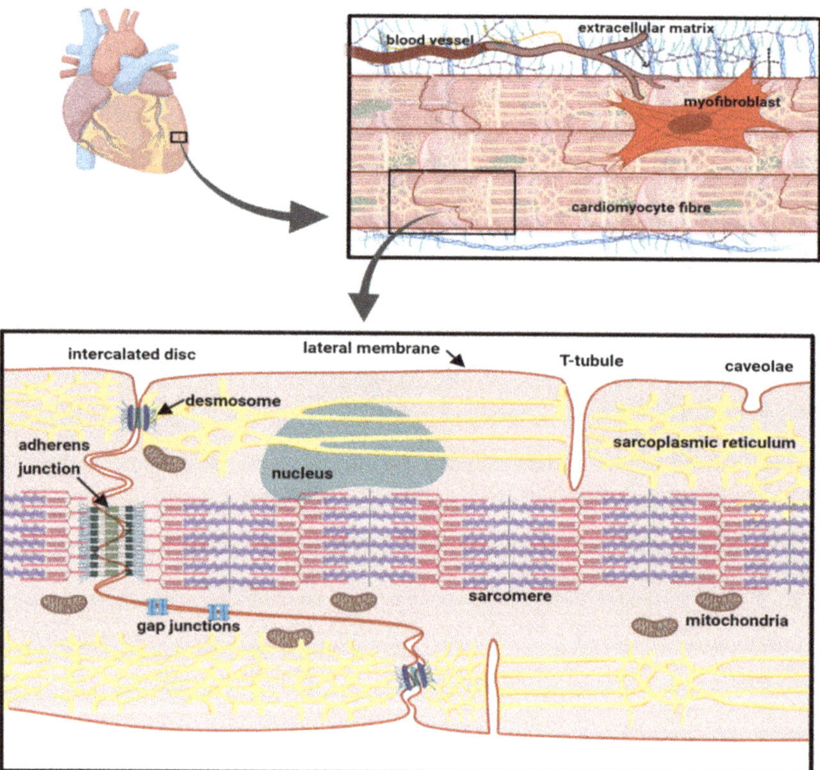

Figure 1. The cardiomyocyte: its anatomical and cellular context. The location of key organelles, membrane compartments and molecular components mentioned in the text are indicated.

1.1. The Nav Channel α-Subunit

All Nav channel α-subunits contain four internally homologous domains (DI-IV). Each domain contains six transmembrane alpha helices (S1–S6) (Figure 2A). Helix S4 of each domain contains positively charged amino acid residues along one face of the helix. The movement of the S4 helices in response to changes in membrane potential is transmitted to helices S5 and S6 of each domain. This leads to the transient opening and subsequent inactivation of the channel pore [8,9]. High-resolution structures obtained by cryogenic electron microscopy (cryo-EM), for the heart-specific Nav1.5 α-subunit, the skeletal muscle channel Nav1.4 and the neuronal channels Nav1.2 and Nav1.7 show that the four domains surround the central pore with four-fold pseudosymmetry. Helices S1–S4 lie on the outer rim of the channel, with helices S5 and S6 from each domain forming the channel pore region [10–14]. This topology is highly conserved between Nav α-subunit isoforms, as illustrated by comparison of the Nav1.7 and Nav1.5 structures (Figure 2B).

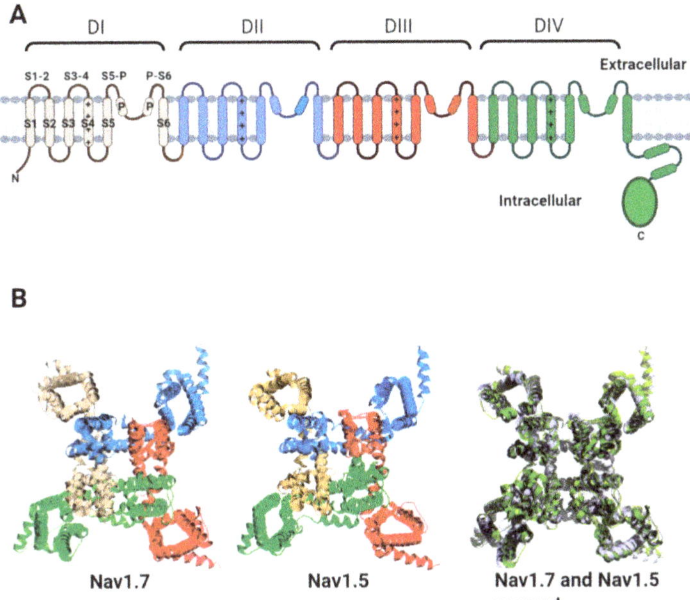

Figure 2. The Nav channel α-subunit. (**A**) Cartoon representation showing internally homologous domains DI-DIV. In DI, the location of transmembrane alpha-helices, S1–S6, the extracellular loops (S1-2; S3-4; S5-P and P-S6) and the re-entrant P helices are indicated. The positive charges on the S4 helices of each domain are indicated. (**B**) Three-dimensional structures of human Nav1.7 (PDB: 6JH8I), rat Nav1.5 (PDB: 6UZ3), and their aligned structures. The channels are viewed from above the plane of the plasma membrane. For Nav1.5 and Nav1.7, the domains DI-DIV are coloured as in (**A**). For aligned structures, Nav1.7 is coloured blue–white and Nav1.5 is coloured pale green.

1.2. The Nav Channel β-Subunits and Their Binding Sites on the α-Subunits

Vertebrate Nav channels are typically associated with one or more β-subunits (Mwt ~30–40 kDa). There are four homologous β-subunit genes (*SCN*1b-4b) encoding subunit proteins β1-β4 respectively. The β-subunits are type I transmembrane proteins consisting of a single extracellular N-terminal V-type immunoglobulin (Ig) domain, connected to a transmembrane alpha-helix by a flexible neck and terminating in a largely disordered intracellular C-terminal region (Figure 3A,B). An alternatively spliced form of β1, known as β1B, is also expressed in the heart. It consists of an Ig domain identical to that of β1, but lacks the transmembrane alpha-helix and is therefore secreted (Figure 3A) [15]. The β1- and β3-subunits show the closest sequence similarity to each other and are more distantly related to β2 and β4 (Figure 3C) [16,17]. The β-subunits have multiple effects on Nav channel gating behaviour that vary between individual β-subunit isoforms. In general terms however, they can increase the peak current density of Nav channels, probably by enhancing trafficking to the plasma membrane [2]. They also shift the voltage ranges over which Nav channel steady-state activation and/or inactivation occur, and in some cases enhance the rates of inactivation and recovery from inactivation [18]. As an illustrative example, the β3-subunit shifts the $V_{\frac{1}{2}}$ for inactivation of Nav1.5 in a depolarising direction: i.e., the voltage at which half the channels are inactivated is displaced to a more positive value compared to the α-subunit alone (Figure 3D) [19–22]. For a cardiomyocyte with a resting potential of about -90 mV [2], this would act to increase the fraction of functional Nav1.5 channels available in the membrane [17].

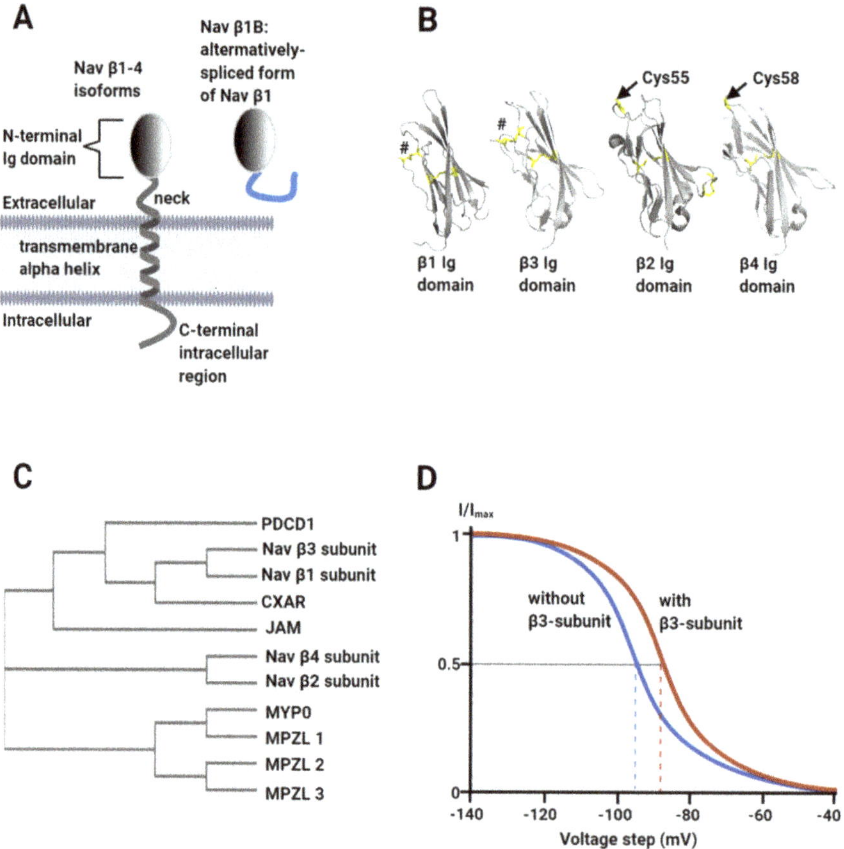

Figure 3. The Nav channel β-subunits. (**A**) Cartoon showing the common structural features of the β-subunits, including the alternatively spliced β1B isoform. (**B**) Atomic-resolution structures for the Ig domains of: β1 (PDB: 6JHI); β2, (PDB: 5FEB); β3 (PDB: 4L1D) and β4 (PDB: 5XAX). The separate disulphide bonds stabilising the N-terminal strands of β1 and β3 are labelled by hashtags and the free, exposed Cys residue on β2 (Cys55) and β4 (Cys58) are as indicated. (**C**) Phylogenetic analysis of Nav channel β-subunits, showing their relationship to members of the Ig domain-containing CAM protein family. PDCD1: programmed cell death protein 1 (https://www.uniprot.org/uniprot/Q15116); CXAR: Coxsackievirus and adenovirus receptor (https://www.uniprot.org/uniprot/P78310); JAM: junctional adhesion molecule 2 (https://www.uniprot.org/uniprot/P57087); MYP0: myelin protein P0 (https://www.uniprot.org/uniprot/P25189); MPZL1: myelin protein zero-like protein (https://www.uniprot.org/uniprot/O95297); MPZL2: myelin protein zero-like protein 2 (https://www.uniprot.org/uniprot/O60487) and MPZL3: myelin protein zero-like protein 3 (https://www.uniprot.org/uniprot/Q6UWV2). The phylogenetic tree was constructed using the ClustalW2 package (https://www.ebi.ac.uk/Tools/phylogeny/simple_phylogeny/). (**D**) Idealised inactivation curves of the Nav1.5 channel in the absence (blue) and the presence (red) of the β3-subunit. The β3-subunit induces a depolarising (rightward) shift of the $V\frac{1}{2}$ of inactivation, as indicated on the diagram by the dotted lines.

The β1-subunit interaction site has been resolved at high resolution for Nav1.2, Nav1.4 and Nav1.7 α-subunits [10–13] and is illustrated for the case of Nav1.7 in Figure 4A,B. The β1-subunit Ig domain makes ionic and hydrogen-bond contacts with the DI, S5-P extracellular loop, the DIII, S1–S2 extracellular loop and the DIV, P-S6 extracellular loop regions (Figures 2 and 4B). Surprisingly however, the Nav1.5 α-subunit structure has revealed some localised, but structurally significant differences between Nav1.5 and the other studied Nav channels [14]. In particular, the Nav1.7 Glu307 residue in the DI, S5-P extracellular loop, is changed in Nav1.5 to an asparagine residue, Asn319. This creates an N-linked glycosylation site that is not present in any other Nav channel isoform. In the Nav1.5 cryo-EM structure, there is electron density around Asn319 that is consistent with a complex N-linked glycan (Figure 4C). It should be noted that the electron density detected in the cryo-EM data only corresponds with two N-acetyl glucosamine residues of the core glycan. The remaining, diverse sugar moieties of the terminal branches are not resolved, presumably due to their inherent flexibility. Thus the N-linked glycan attached to Nav1.5, Asn319 extends further than the resolved electron density and would certainly be bulky enough to occlude the binding site for the β1 Ig domain [12]. Moreover, the specific orientation of a second N-linked glycan attached to Nav1.5 residue Asn1390 will probably also interfere with the binding of the β1 Ig domain (Figure 4C). Hence, it seems likely that in vivo, although β1 may still be associated with the Nav1.5 DIII voltage sensing domain via its transmembrane alpha-helix, its Ig domain will not be able to bind to the Nav1.5 α-subunit.

Based on biochemical and electrophysiological data, it is probable that the β3-subunit transmembrane alpha-helix also binds to Nav1.5 DIII voltage sensing domain [19–21]. Yet, there is evidence that it may bind closer to Nav1.5 DIII helix S3 rather than to the binding site for the β1 transmembrane region on the DIII helix S2 [19,21]. If so, then a given Nav1.5 α-subunit may be able to bind simultaneously to β1 and β3-subunits and there is indeed electrophysiological evidence to support this idea [21,23].

In contrast to β1 and β3, which bind to the α-subunit non-covalently, the β2-subunit binds to Nav1.7 covalently via a disulphide bond between a cysteine on the Ig domain (Cys55) and a corresponding cysteine (Cys895) on the α-subunit DII S5-P extracellular loop (Figure 4D) [12,24]. Neither the transmembrane alpha-helix nor the intracellular region of β2 is resolved in the published structure, indicating that both must be unconstrained in this purified complex [12]. As with β2, the β4-subunit Ig domain also contains a cysteine (Cys58) that can form a disulphide bond to the free cysteine on the α-subunit DII, S5-P site [25]. It is therefore presumed that the β2 and β4-subunit Ig domains covalently bind to the same or largely overlapping site on most Nav channel α-subunits [26,27]. Oddly however, the putative β2- and β4- subunit binding-site in Nav1.5 again shows important sequence differences from that on other Nav channels. Most notably, the residue equivalent to Cys895 of Nav1.7 is changed to leucine in Nav1.5 (Leu869) (Figure 4D–F). Since there are no other accessible free cysteines on the Nav1.5 extracellular surface, it will be impossible for either the β2- or the β4-subunit Ig domains to covalently bind Nav1.5 as they do to Nav1.7. Furthermore, the amino acid residues clustered around Nav1.7 Cys895 and which, in Nav1.7 provide additional contacts with the β2 Ig domain, are substantially different in Nav1.5 (Figure 4F).

Taken together, this evidence suggests that the Ig domains of all four β-subunits will be unable to bind to Nav1.5 directly, although the β-subunit transmembrane and intracellular regions may still do so. As a result, the Ig domains will be free to explore a greatly extended volume space above and around the Nav1.5 channel than β-subunits attached to most other Nav channels. What then are the likely functional consequences of this difference?

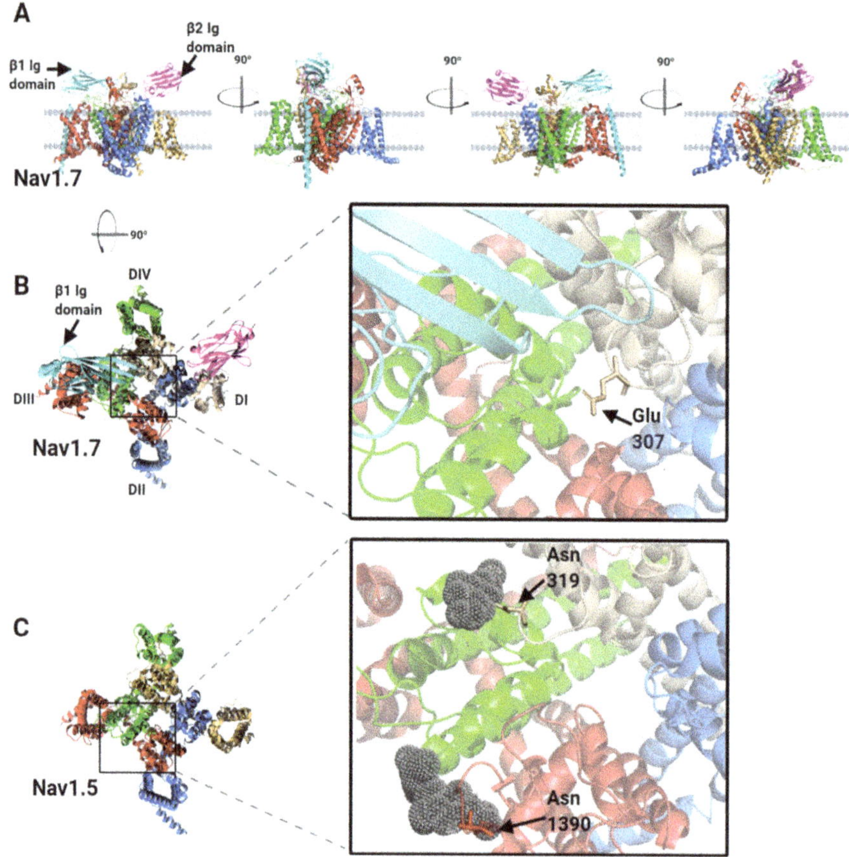

Figure 4. *Cont.*

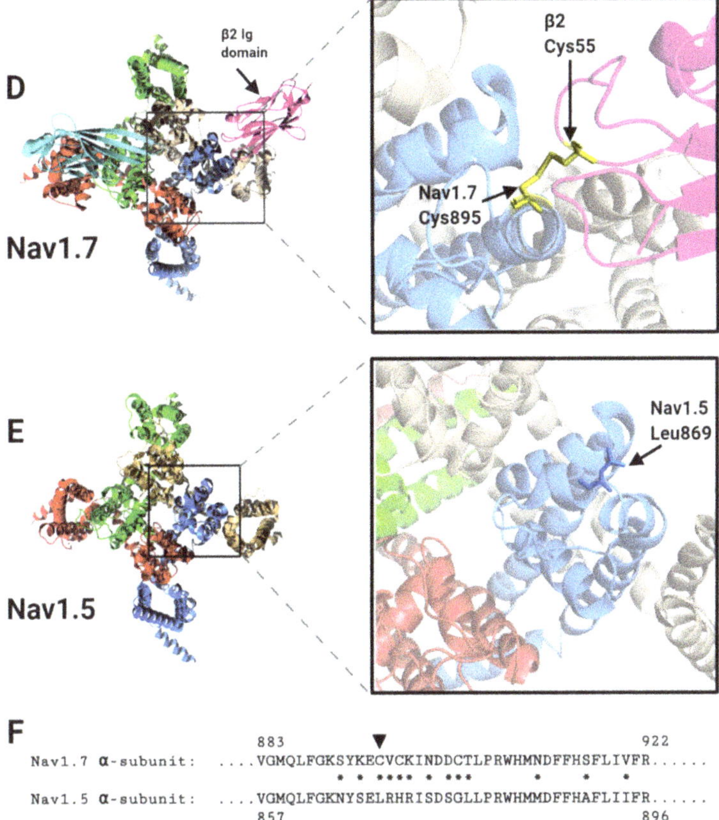

Figure 4. The binding sites for β1 and β2 on the Nav1.7 α-subunit and its comparison with Nav1.5. (**A**) Side views, each with 90° rotation, of the Nav1.7 α-subunit, with associated β1 and β2 -subunits. (**B**) Top view of the Nav1.7 α-subunit, with the β1 Ig domain binding site on the DI, DIII and DIV extracellular loops highlighted. (**C**) Top view of the equivalent region of Nav1.5 α-subunit. Resolved electron density corresponding to the N-linked sugar residues mentioned in the text are shown in grey dots. (**D**) Top view of the Nav1.7 α-subunit with the β2 Ig domain binding-site on DII extracellular loop highlighted. (**E**) Top view of the equivalent region of the Nav1.5 α-subunit. (**F**) Sequence alignment of human Nav1.7 and Nav1.5 α-subunits around the DII β2 binding-site. Amino acid differences between the two sequences are indicated by asterisks. The position of the Cys895 residue in Nav1.7, noted in the text, is indicated with an arrowhead.

2. The Nav Channel β-Subunits as Cell-Adhesion Molecules

Phylogenetic analysis indicates that all four of the β-subunits are closely related to the myelin P0 family of cell-adhesion molecules (CAMs). Indeed, the β2 and β4-subunit sequences are more closely related to other myelin P0-like proteins, MPZL1-3, than they are to either β1- or β3-subunits [28–30] (Figure 3C). The β1-subunit can bind in *trans* both to itself and to other CAMs such as neurofascins and contactins [31]. The β2-subunit can bind in *trans* to extracellular laminin [32]. The β3-subunit can bind homophilically in *cis* [20,33] and heterophilically in *trans* with neurofascins [34]. The β4-subunits can bind homophilically both in *cis* and *trans* [35]. In all cases, these interactions occur via the Ig domains [36]. Hence, the Nav β-subunits can be considered CAMs in their own right; albeit with the additional property of Nav channel modulation.

2.1. The Nav1.5-Associated β1-Subunit as a CAM at the Intercalated Disc: Its Role in Ephaptic Conduction

The intercalated disc is formed from the juxtaposition of the sarcolemma membranes from adjacent cardiomyocytes and enables the cells to be mechanically and electrically coupled together [37]. It has a complex inter-digitated 'plicate' structure that ensures the inter-cellular connections are both close and continuous. Within the intercalated disc, there are distinct and structurally diverse domains each with their own specialised functions. For example, tight physical coupling at the desmosome and adherens junction help to transmit mechanical force between adjacent cells (Figure 1) [38,39]. Interspersed between these structures lie the gap junction plaques [40]. These domains contain a high local density of the protein connexin. Ventricular cardiomyocytes mostly express the connexin-43 isoform, although other isoforms are also present [41]. Connexins co-assemble on each membrane to form a symmetrical hexameric pore-containing hemichannel. Two hemichannels - one from each apposing membrane - associate in *trans* to form the functional gap junction and bind the two membranes together, so that they are no more than 2–4 nm apart, a distance that provides a tight inter-membrane seal [42]. Furthermore, a typical plaque contains several hundred closely packed gap junctions, ensuring the almost complete exclusion of other proteins from this region [41]. The central connexin pore is large enough to permit the passage of ions and small molecules [43]. Hence, the common view that electrical coupling between cardiomyocytes occurs predominantly by an electrotonic spreading mechanism in which charged ions flow passively from the cytoplasm of the 'upstream', depolarised cell to the cytoplasm of its neighbouring, 'downstream' quiescent cell via their shared gap junctions (Figure 5A) [44–46].

Surrounding and abutting the gap junction plaques lies the perinexus [47,48]. Unlike the gap junction, the perinexal membranes from adjacent cells do not interact directly, but enclose a restricted, inter-perinexal space, about 20 nm wide and extending away from the gap junction plaques for about 200 nm [48]. The perinexal membranes display a distinctive molecular composition. In particular, they contain a high local density of Nav1.5 channels that form multi-molecular two-dimensional clusters on the membrane surfaces [49]. This clustering may partly reflect an inherent tendency of Nav1.5 channels to self-associate [20]. But Nav1.5 channels are additionally stabilised by interactions with connexin-43 hemi-channels and with multi-modular cytosolic scaffolding proteins such as ankyrin-G, ZO-1 and 14-3-3 [50–52]. These interactions not only ensure the accumulation of Nav1.5, but they can also modify Nav1.5 activity. For example, ankyrin-G acts as a coordinating signalling hub, functionally connecting Nav1.5 gating with upstream kinase and phosphatase enzymes and down-stream cytoskeletal proteins [53]. In addition, 14-3-3 promotes co-operative gating behaviour between Nav1.5 α-subunits [54].

The clustering of Nav1.5 channels on the perinexal membranes, with the potential for synchronous depolarisation, ensure that activated channels on one membrane can withdraw enough sodium ions from the restricted inter-perinexal space, so that the reduced density of positive charge depolarises Nav channels on the perinexal membrane from the adjacent cell. This is the concept of ephaptic conduction, in which electrical communication between cardiomyocytes is achieved without the direct movement of sodium ions from one cell to the other (Figure 5B) [46,47,55–58]. Interestingly, perinexal Nav1.5 channels co-localise with the inwardly rectifying potassium channel Kir2.1 [59,60]. There is structural evidence that the C-terminal domains of Nav1.5 and Kir2.1 co-assemble on the cytoskeletal protein SAP97, and thereby form a functionally linked, macromolecular complex [61,62]. The Kir2.1 channel exhibits a relatively high potassium conductance when the cardiomyocyte is at the resting potential. Initially therefore, the movement of potassium ions into the perinexal space will buffer the local change in electrical potential when the Nav1.5 channels first begin to open. However, its inwardly rectifying property leads to a reduced conductance during the action potential plateau phase [63]. At this point, the entry of positively charged potassium ions into the perinexal space is minimised, just as the Nav1.5 channels are maximally depolarising and removing positively charged sodium ions. Hence, the Kir2.1 channel may act synergistically with Nav1.5 to finely regulate the rate and extent of ephaptic conduction [59].

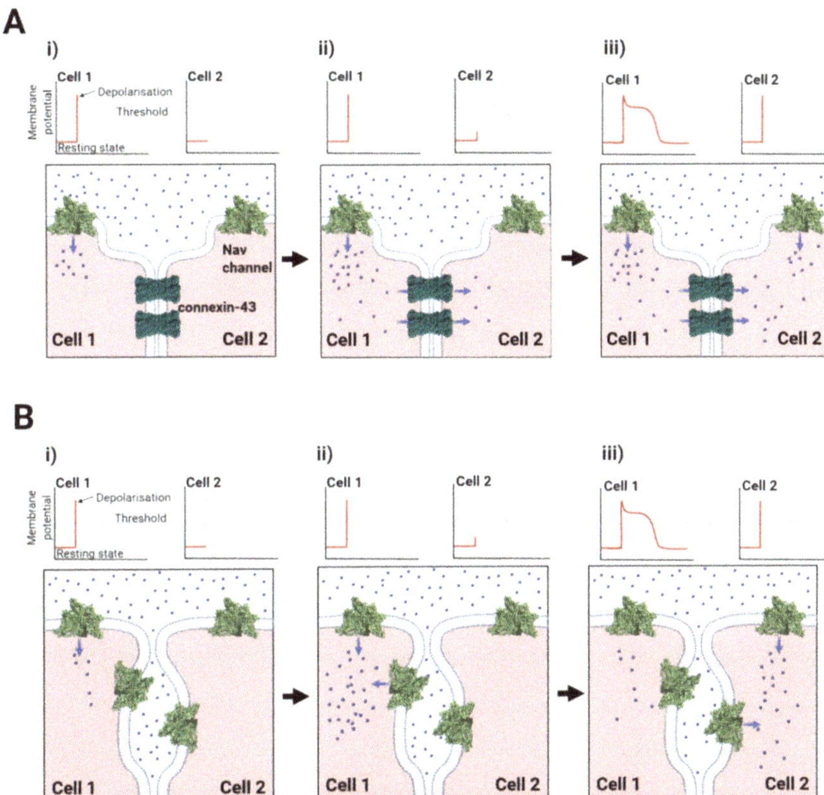

Figure 5. Comparison of electrotonic and ephaptic conduction mechanisms. (**Ai**); In electrotonic conduction, depolarisation of cell 1 by an action potential generates a membrane potential across the gap junction comprising connexin molecules. (**Aii**); The resulting current flow from active cell 1 to quiescent cell 2 causes its membrane to depolarise. (**Aiii**); When this changing membrane potential rises above the threshold for Nav channel opening, an action potential is initiated in cell 2 and the excitation wave is propagated. Note, the involvement of direct charge transfer between successive cells in this process, which involves an *ohmic* transfer of charge. (**Bi**); In ephaptic conduction, Nav channel excitation depolarises cell 1. (**Bii**); As a result, Nav channels in the perinexal membrane of cell 1 become depolarised and there is a removal of sodium ions (and hence a net removal of positive charge) from the ephaptic space separating cells 1 and 2. (**Biii**); The negative change in electrostatic potential within the restricted ephaptic space is enough to depolarise the transmembrane potential in cell 2, causing the activation of its Nav channels and propagation of the excitation wave. Note the absence of direct charge transfer between successive cells in this process, which involves an *electrostatic* transfer of charge. For clarity, only connexin-43 gap junction channels and Nav channels are shown, and in the action potential profiles, the Nav channel threshold has been displaced upwards.

Triggering ephaptic conduction would constitute a distinct non-canonical function of Nav channels in promoting cell-to-cell excitation, as opposed to the more familiar action potential conduction within the same cell. It could complement, or in some pathological circumstances, potentially replace electrotonic conduction mediated by gap-junctions [55,58]. Conversely, their disruption following pathological inflammatory or fibrotic change might compromise cell-cell conduction [64]. This could be in addition to disruptions in connexin-mediated cell-cell coupling [65]. It is in this context, that we suggest the unique structural features of Nav1.5 and its associated β-subunits should best be interpreted.

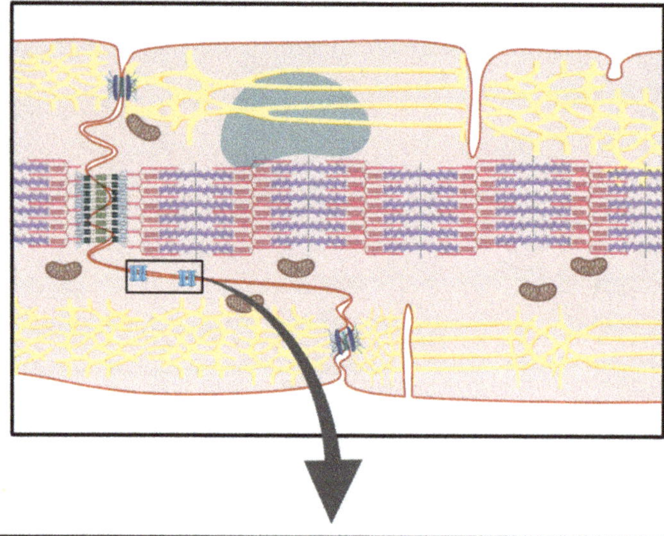

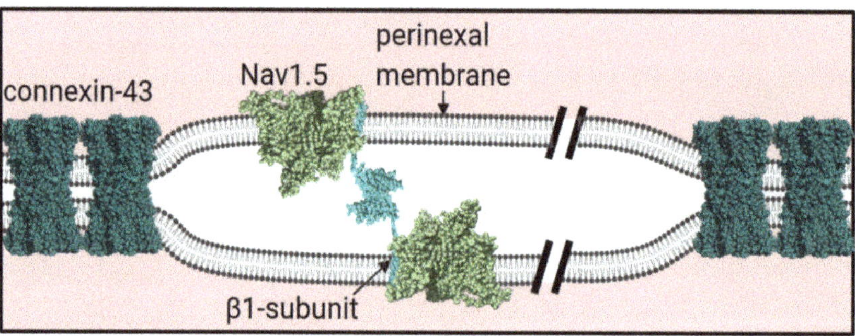

Figure 6. The proposed role for the β1-subunits in stabilising Nav1.5 channels on both apposing perinexal membranes, whilst also maintaining the necessary width between perinexal membranes to ensure efficient ephaptic conduction.

The perinexal Nav1.5 channels are associated with the β1-subunit. Recently, *Veeraraghavan, et al.*, have shown that the perinexal membranes are associated with each other by *trans* homophilic binding between apposing β1 Ig domains [66]. However, if the β1 Ig domain is attached to Nav1.5 in the way that it is to Nav1.7, then it would only protrude about 4–5 nm above the membrane surface (Figure 4A). In that case, the Ig domains would not be able to cover the required 20 nm distance between perinexal membranes [47]. If on the other hand, the β1 Ig domain is not anchored onto the Nav1.5 α-subunit, then the two β1-subunits from apposing membranes are freer to extend their flexible necks and bridge this gap, forming *trans* cell-adhesion contacts (Figure 6). We suggest that the unique sequence differences in the DI and DIV extracellular loop domains of the Nav1.5 α-subunit compared to other Nav channels (Section 1.2, above) are specific adaptations that prevent Nav1.5 binding to the β1 Ig domain and so facilitate this specialised cell-adhesion behaviour of the β1-subunit. In consequence, the *trans*-mediated β1 Ig domain interactions help define the dimensions of the perinexal space by constraining its width to no more than about 20 nm. Mathematical modelling indicates that a gap of this order is optimum for ephaptic conduction [56].

The two interacting β1 Ig domains form an extended antiparallel contact surface running between residues 66 and 86 (Figure 7A,B) [66,67]. In ventricular cardiomyocytes, a peptide mimetic of this binding site competitively inhibited *trans* binding between the two β1 Ig-domain. This induced a widening of the perinexal space, with a concomitant reduction in perinexal sodium current that precipitated arrhythmogenesis [66]. The clinical relevance of this work is supported by studies on patients with atrial fibrillation that show a correlation between arrhythmic severity and excessive widening of the atrial perinexal membranes [68]. Moreover, there are two independent *SCN*1b mutations within or close to this region that are associated with Brugada syndrome: a charge-neutralising Arg to His change at residue 85 and a charge-neutralising Glu to Gln mutation at residue 87 (Figure 7B) [69,70]. Taken together, these multiple structural, anatomical, pathological and experimental data sets not only support a role for ephaptic conduction between cardiomyocytes, but also implicate the structure, geometry and *trans*-binding behaviour of the β1-subunit as specific adaptations that are critical for this mechanism.

When the cardiomyocytes are at rest, the inter-perinexal space is largely enclosed, such that it is minimally influenced by its surrounding extracellular fluid. This too should increase the ability of activated Nav1.5 channels to remove enough positive charge to induce ephaptic conduction. Nevertheless, there must be some way for the electrolyte composition of the perinexal space to be reset between excitation events. This could be facilitated by geometrical changes produced by the cardiomyocyte contractions if they transiently enhance the accessibility of the perinexal space to the extracellular milieu. Real-time in vivo imaging of ventricular cardiomyocytes has identified surprisingly large flexing movements in the perinexus during the propagation of cell-to-cell contraction [71]. A consequence of these cellular and membrane movements will be the rhythmic stretching and relaxing of the *trans*-associated β1-subunits. Here, the disordered and spring-like neck region of the β1-subunit, connecting its rigid Ig domain to its rigid transmembrane alpha-helix might enable some strain-absorbing movement of the β1-subunit. However, as noted above, the β1 peptide mimetic can not only gain access to the perinexal space, it can also disrupt the *trans*-binding between β1 Ig domains [66]. This implies that the *trans* binding must—at least to some degree—be dynamic, perhaps reflecting a periodic dissociation and rebinding during the stretching cycle.

The β1-subunit can bind to cytosolic ankyrin-G and ankyrin-B via its intracellular region. But this interaction is abolished if a critical intracellular tyrosine residue, Y181, is phosphorylated by Fyn kinase [72,73]. In cardiomyocytes, the tyrosine-phosphorylated form of β1 is present only at the intercalated disc, where it interacts with Nav1.5 and the CAM cadherin, but does not bind ankyrin-G. The non-phosphorylated β1-subunit is present on the lateral membranes [74]. Of course, there could be many functional reasons for this pattern. For example, phosphorylation may be a targeting signal to direct β1-subunits to the intercalated disc [15]. Alternatively, the β1-subunit may be actively phosphorylated at the intercalated disc in response to changes in membrane stretching or other physiological signals. This would change the proportion of β1 that could connect to the cytoskeleton via ankyrin-G and thus modify cell-adhesion, perhaps even during a contraction cycle. One may speculate that the β1-mediated *trans* contacts could be further modulated via the alternatively spliced β1B isoform (Figure 3A) [15]. Since this isoform is secreted, it could in theory interrupt and fine-tune the *trans*-binding, if present in the perinexal space. There is as yet no direct evidence for this proposal, but a mutation in the β1B isoform is linked to Brugada syndrome [75].

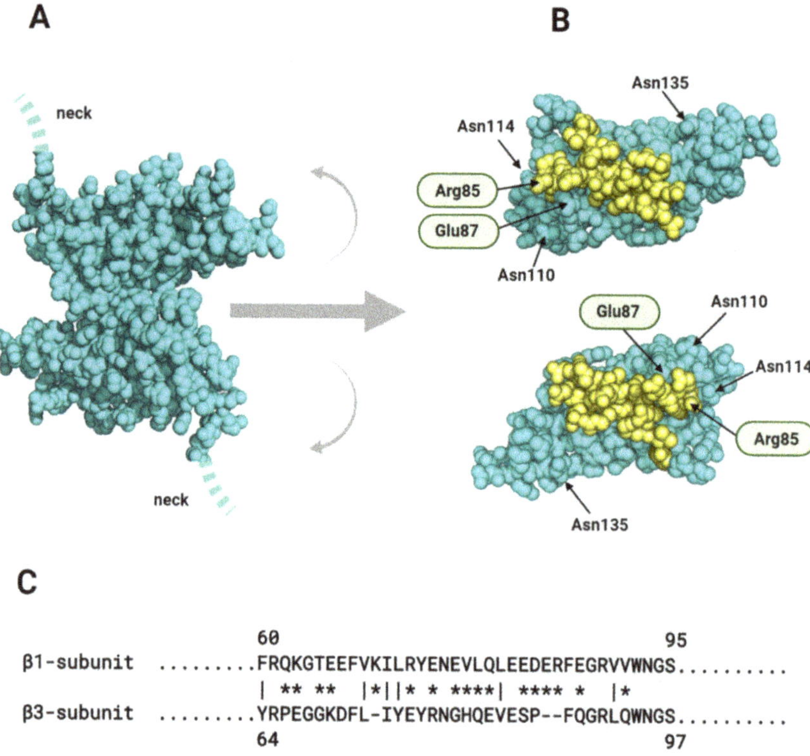

Figure 7. The proposed contact surface between the trans-interacting β1 Ig domains. (**A**) The antiparallel arrangement of β1 Ig domains. (**B**) the two β1 Ig domains 'peeled back' to reveal the contact surface proposed by [66]. The location of two Brugada mutations (Arg85 and Glu87), mentioned in the text are indicated. Potential N-linked glycosylation sites, Asn110, Asn114 and Asn135 are indicated. (**C**) Sequence alignment between β1 and β3 Ig domains within the proposed contact surface, showing low sequence identity. Non-conservative changes are shown with asterisks and conservative changes with vertical lines.

2.2. Do Other Nav Channel β-Subunits Facilitate Ephaptic Conduction?

Although the β3 and β1-subunits show the closest overall sequence similarity (Figure 3C), there are localised sequence differences between them that could suggest functional specialisation. In particular, the sequence of the β1 Ig domain *trans* cell-adhesion binding site is not conserved in β3 (Figure 7C) [76] and the evidence that β3 can act as a homophilic *trans* CAM is mixed [76,77]. We therefore suggest that unlike β1, the β3-subunit may not play a major role in stabilising the perinexal space by *trans*-mediated cell-adhesion. On the other hand, when expressed alone in HEK293 cells, the β3-subunits bind homophilically in *cis*, using their Ig domains [19,20,33]. Super-resolution imaging experiments show that when co-expressed with Nav1.5, the β3-subunit affects the relative geometry between Nav1.5 channel α-subunits, possibly by promoting particular orientations of individual Nav1.5 α-subunit dimers within larger clusters [20]. This is consistent with a *cis*-mediated β3-subunit effect. The *cis*-interacting β3-subunit Ig domains could cross-link Nav1.5 channels within each perinexal membrane and thus contribute lateral stability to the channel clusters whilst the β1-subunit provides structural stability between apposing perinexal membranes.

Binding of the β2-subunit to Nav1.5 cannot be detected by immunoprecipitation, when tested in transfected HEK-cells [14]. Yet the β2-subunit can be immunoprecipitated together with Nav1.5 from heart tissue [78] and mutations in the β2 cytosolic region compromise the trafficking efficiency of Nav1.5 in cardiomyocytes [2]. Hence, additional cardiomyocyte-specific protein contacts are likely to be required to stabilise β2-binding to Nav1.5 in its normal physiological context. Interestingly, the Ig domains of the β1 and β2-subunits can interact in *trans*, but the association requires the presence of a sequence within the β2-subunit cytoplasmic region that contains a putative casein kinase II phosphorylation site [31]. This raises the possibility that *trans* heterophilic interactions between β1 and β2 Ig domains may occur across the perinexal space, yet be regulated by signal transduction events acting on the cytosolic face of the membranes [2].

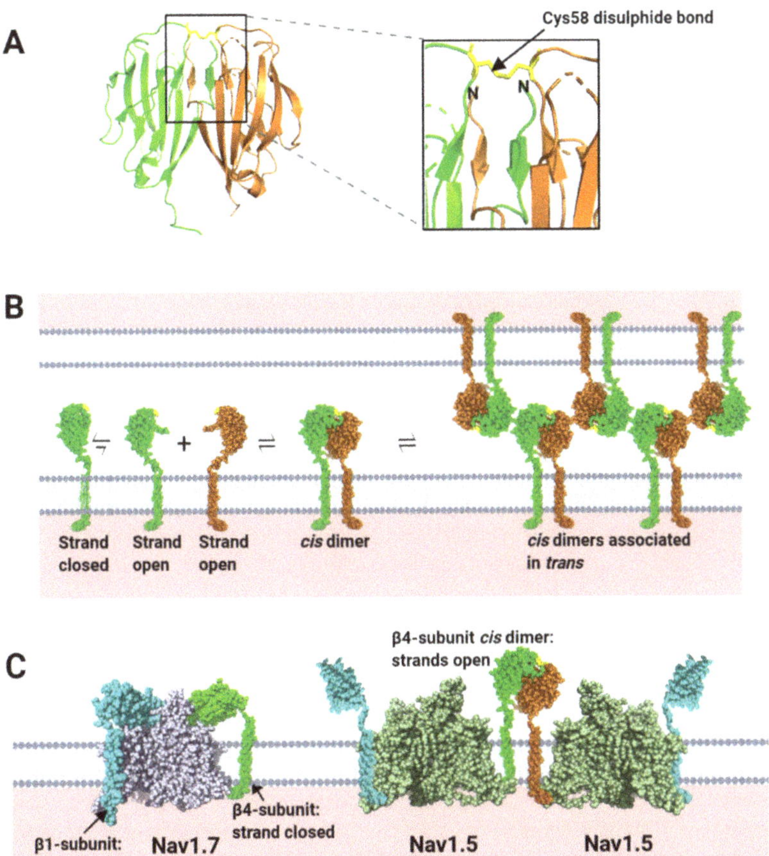

Figure 8. The β4-subunit Ig domain can associate in both *cis* and *trans*. (**A**) The disulphide-bonded, *cis*-interacting, dimeric β4-subunit Ig domain (PDB: 5XAW) with the Cys58 disulphide and the strand swap N-terminal regions highlighted. (**B**) Proposed model for *trans*-interacting, β4-subunit Ig domain *cis*-dimers, based on the model discussed in the text [79]. (**C**) Proposed association of the β4- and β1-subunits on Nav1.7 and Nav1.5 α-subunits. Note, to clarify the structural details in (**A**) and the cartoon models in (**B**) and (**C**), the two identical β4 Ig domains and β4-subunits are coloured differently, in green and orange.

The homophilic cell-adhesion contacts of the β4 Ig domain have been studied using crystallographic analysis, site-specific cross-linking with unnatural amino acids and cell-adhesion assays. The β4 Ig domain can be isolated as a dimer, stabilised by two striking features. Firstly, the presence of an inter-subunit disulphide bond between the Cys58 residue on each Ig domain. Secondly, a reciprocal strand-swap interaction, in which the first seven N-terminal amino-acid residues from one Ig domain interact with residues on the partner Ig domain (Figure 8A) [79]. N-terminal strand-swapping is a known feature of several *trans*-mediated CAMs that contain Ig domains, including cadherins in the adherens junctions and desmogleins in the desmosomes [80–82]. In the case of β4 however, the strand-swapped, disulphide-bonded Ig domain dimer is proposed to assemble on the same membrane in a *cis*-interaction. In this process, the N-terminal residues that undergo swapping must first undock from their binding site within the 'closed' monomer, to generate an 'open' monomer, which is then capable of *cis*-dimerisation (Figure 8B). A β4 Ig domain in which the Cys58 residue was changed to alanine, crystallised as a monomer in the asymmetric unit and the N-terminal strand was not resolved [26], suggesting that this conformation might correspond to the monomeric 'open' state. It has been further proposed that at the cell-surface, the β4 *cis* dimers, then interact in *trans* to promote cell-adhesion (Figure 8B) [35,79]. It is interesting to note that neither the β1-subunit, nor the β3-subunit Ig domains can form similar strand-swap dimers, because their N-terminal strands are covalently locked in place via an intramolecular disulphide bond [12,33,76] (Figure 3B).

These structural insights have an important implication. If β4 is paired with most Nav channel α-subunits, its Cys58 residue will form a disulphide bond with the free cysteine on the α-subunit DII S5-P extracellular loop site [25] (e.g., Nav1.7; Figures 4D and 8C). But this would inhibit formation of the Cys58-mediated β4 *cis* dimer (Figure 8A,B). On the other hand, the Nav1.5 α-subunit lacks a free DII, S5-P cysteine. So, it is not possible for Nav1.5 to covalently bind the β4-subunit Ig domain (Section 1.2.). Nevertheless, the interaction between the β4-subunit and Nav1.5 is stable enough to be detected by immunoprecipitation when co-expressed in HEK293 cells [83]. Presumably, β4 must still bind to Nav1.5 via other sites such as the transmembrane alpha-helix and/or the cytoplasmic region. This will leave any Nav1.5-associated, β4-subunit Ig domains free to form disulphide-bonded, homophilic dimers (Figure 8C). Thus, potentially, the β4-subunit could facilitate not only *cis*- but also *trans*-mediated Nav1.5 cross-linking (Figure 8B,C). The overall dimensions of the β4 subunit are similar to β1 [17]. The β4 and β2-subunits can be detected very close to - but distinct from - the intercalated disc gap junctions [5], suggesting a likely perinexal location. It is tempting to speculate that the β4 (and probably β2) subunits could help stabilise both the width of the perinexal space and the clustering of Nav1.5 within this region by facilitating extensive *cis* and *trans*-mediated associations.

2.3. Nav Channels and β-Subunits on the Lateral Membrane: A Role in Mechanosensing?

A major fraction of the Nav1.5 channels on the lateral membrane are sequestered into lipid rafts and caveolae [84,85]. The caveolae are cholesterol-rich microdomains stabilised by the intrinsic scaffold membrane protein caveolin and the peripheral, cytosolic protein cavin [86]. Palmitoylation is commonly thought to act as a targeting signal that directs proteins into these structures [87]. Interestingly, the transmembrane alpha-helix of the β1 and β3-subunits (but not the β2 and β4-subunits) contains a juxtamembrane cysteine residue, located at the cytoplasmic interface that fits the consensus for a palmitoylation site [16,17]. Recent work confirms that β1 is indeed palmitoylated at this residue (Cys162). However, mutational abolition of the Cys162 residue does not prevent β1 targeting to detergent-resistant membrane fractions (assumed to correspond to lipid rafts), although it does reduce the steady-state level of the β1-subunit at the plasma membrane [88]. Within the caveolae, the Nav1.5 and β-subunit complexes are co-clustered with both Kir2.1 and the Kv4.2/4.3 channels responsible for the transient outward potassium current [84,89,90]. Surprisingly, the Nav channel β1-subunit binds to and regulates the Kv4.2/4.3 channels [91]. This opens the potential for mutual control of both channels via Nav β1 - especially so on the restricted and crowded caveolar membrane. The caveolae also sequester integrins, required for cell-adhesion to extracellular matrix proteins such as collagen and fibronectin, together with the integrin-activated signalling enzyme, Fyn kinase [86,92]. However, since the caveolae are highly invaginated, rosette-like clusters, it is not clear whether the sequestered integrins can bind extracellular matrix proteins under these conditions; neither is it clear whether the sequestered ion channels can be active [93]. Yet there is evidence that caveolin can regulate integrin activity [94] and in turn, both integrins and caveolin can modulate Nav1.5, Kir2.1 and Kv4.2/4.3 channels [95–97].

A possible explanation for these observations is that caveolae act as sensors for mechanical stress and respond dynamically to changes in cell deformation [98]. In particular, the cell stretching initiates a rapid dissociation of the cavin coat and a 'flattening' of the caveolae membrane. This mechanism increases the surface area of the plasma membrane and helps buffer against stretching forces [99,100]. A further consequence is that the integrins, ion channels and signal transduction enzymes are repartitioned into the plasma membrane, but probably remain as a macromolecular complex. This translocation is followed by an increased integrin-stimulated ERK signalling, which is caveolin-dependent [93]. Since ERK is a down-stream target of activated Fyn kinase [101], this suggests that the translocation of the integrin/ion channel complex into the wider plasma membrane can enhance integrin-dependent Fyn-kinase activation, perhaps by enabling greater access of the integrins to their extracellular matrix ligands. Activated Fyn kinase also phosphorylates Nav1.5, inducing a depolarising shift in the $V_{1/2}$ of steady-state inactivation, thus increasing the fraction of functionally active Nav1.5 channels on the membrane [102]. Furthermore, Nav1.5 is a mechanosensor and the membrane flexing itself will affect channel gating. In particular, the Nav1.5 channel responds to physiological levels of mechanical membrane stress with a hyperpolarising shift to the $V_{1/2}$ of steady-state inactivation [103,104]. Remarkably, the β1-subunit further amplifies this hyperpolarising shift [105]. The combination of increased Nav1.5 excitability via integrin-activated Fyn-kinase signalling, with the opposite mechanosensitive response of Nav1.5 and its β1-subunit may act to finely balance the need to integrate different cell-adhesion signals, whilst also protecting against excessive mechanical stress over a large voltage range. The β3-subunit enhances the mechano-induced acceleration of both activation and inactivation kinetics, but unlike β1, it does not affect the mechanosensitive shifts in steady-state parameters [105]. The structural explanation for these differences between β1 and β3-subunits and indeed their functional significance, is not clear.

The β-subunits exposed on the lateral membrane can interact with extracellular matrix proteins. For example, the β2-subunit can bind in *trans* to laminin, a known component of the heart extracellular matrix [32]. But there is relatively little β2-subunit exposed on the cardiomyocyte lateral membrane [5]. Another possibility is that β1-subunits exposed on the surface of the lateral membrane may interact homophilically with β1-subunits and other CAMs on adjacent cell types [31,34]. For example, during cardiac fibrosis, there is an enhanced proliferation of myofibroblasts that express several different Nav channels including Nav1.5, together with the β1-subunit [106]. Myofibroblasts can form aberrant electrical connections to the cardiomyocytes, which likely adds additional and potentially pro-arrhythmic membrane capacitance [107]. Whether the contacts between cardiomyocytes and myofibroblasts involve *trans*-cell adhesion contacts between β1-subunits remains to be determined.

2.4. Nav β-Subunits and Neuronal Channels in the T-Tubules

The neuronal Nav channels localised within the T-tubules are predominantly associated with the β1 and β3-subunits [5]. These Nav α-subunits all lack a glycosylation site in their DI, S5-P extracellular loop, equivalent to Asn319 in Nav1.5 [9]. Hence, it is likely that the β1 and β3 Ig domains will bind onto these Nav α-subunits in a manner more like Nav1.7 than Nav1.5 (Figure 4A,B). As the T-tubule width is significantly greater than 20 nm (and is often more than 100 nm) [108], any direct *trans*-mediated binding of β1 Ig domains between opposite membrane faces of the T-tubules is unlikely. There is evidence that CAMs such as laminin can enter the lumen of large-diameter T-tubules [109]. So, cell-adhesion-type functions for T-tubular β-subunits cannot be ruled out. In mice, deletion of the *SCN1b* gene perturbs the interaction between T-tubular Nav channels and NCX complexes, with a consequent dysregulation of calcium homeostasis, suggesting a structural role to facilitate excitation-contraction coupling [110].

3. Conclusions and Unsolved Problems

On the cardiomyocyte membranes, the Nav channels form heterogeneous, multi-component macromolecular clusters, rather than remain as isolated molecules [111]. There is no necessary requirement for every Nav α-subunit to have an identical stoichiometry with any associated β-subunit [17]. Examples from other ion-channels show that the behaviour of membrane-bound clusters can change depending on variations in subunit ratios [112]. In such assemblies, individual protein components can have more than one function, depending on the physiological context. Although originally identified by their direct effects on channel gating, it is now clear that the Nav β-subunits extend their functions to include cell-adhesion and mechano-sensing and in doing so, raise further questions:

3.1. Evolutionary Relationship between Nav β-Subunits and Other CAMs

The Ig domain superfamily has deep evolutionary roots that pre-date the divergence of vertebrate and invertebrate lineages [113]. Yet the Nav channel β-subunits have only been discovered in vertebrate genomes, where they cluster together with members of the Ig domain-containing CAM family [30]. This close evolutionary relationship raises the intriguing possibility that homologues such as the MPZL1-3 group of proteins (Figure 3C), might act as additional Nav channel modulators. At least some of these proteins are expressed in heart tissue [114].

Although lacking β-subunits, invertebrate Nav channels do possess associated proteins that modulate gating and trafficking behaviour of their Nav channels. The best characterised are members of the TipE family [115]. However, these proteins show no sequence or structural similarity to vertebrate β-subunits and must have independently evolved their Nav channel-modulating behaviour. It will be interesting to see whether the TipE proteins can act as CAMs, or whether cell-adhesion is a unique feature of the vertebrate β-subunits.

3.2. The Biophysics of Nav β-Subunit Cell-Adhesion

We currently lack a quantitative understanding of the *trans*-mediated binding events facilitated by the β-subunits. For example, it would be interesting to know if the contacts between individual β1 Ig domains at the perinexus are strong and stable or individually weak enough to dissociate and rebind rapidly. The latter case might be more likely given the dynamic nature of membrane movements at the intercalated disc during the contraction, relaxation cycle [71]. The application of new biophysical techniques such as atomic force microscopy and traction force microscopy [116], combined with more traditional biochemical and molecular genetic techniques will be needed to address these questions.

3.3. The Role of N-Linked Glycosylation

Membrane proteins are generally N-linked glycosylated, with complex, branching sugar residues, often tipped with sialic acid moieties [117]. The role of N-linked glycosylation in the trafficking of Nav channels, including Nav1.5 - is well-established [118]. There is also evidence that the negatively charged sialic acids on N-linked glycans of Nav channel α and β-subunits can modulate channel gating [119]. In addition, the relatively large and bulky N-linked glycans can potentially modulate the strength and even the possibility of protein-protein interactions occurring. A good example is described above for the case of the β1 Ig domain binding to Nav1.5, and the likely role of the glycosylated Asn319 residue in preventing binding of the β1 Ig domain (Section 1.2, Figure 4C). Another example is in the model proposed for β1- *trans* cell-adhesion. Here, the putative *trans*-binding motif on the β1 Ig domain surface is surrounded by four of its five potential N-linked glycosylation sites (Section 2.1, Figure 7B). Could the strength of this interaction be fine-tuned by for example, developmentally regulated changes in the nature and extent of N-linked glycosylation?

3.4. Ephaptic Conduction in the Heart and Elsewhere

In cardiomyocytes, ephaptic conduction occurs in close association with gap junction structures mediating electrotonic conduction (Section 2.1, Figure 5), suggesting that both processes occur to relative extents, that might vary under different conditions [120]. It is likely that there are other biological situations where the necessary conditions for ephaptic conduction apply. Potential examples include the repetitive firing that occur in neuroendocrine supraoptic nucleus neurones [121,122] and the escape reflex triggered by activation of the goldfish Mauthner neurone [123]. Interestingly, the R85H mutation in the β1 Ig domain, that compromises ephaptic conduction between cardiomyocytes [66] (Section 2.1, Figure 7B), also predisposes to epilepsy [124], perhaps hinting at a similar role in neurones.

3.5. Clinical Implications

Assuming that electrical signalling between cardiomyocytes occurs both by electrotonic and ephaptic mechanisms, then a drug that inhibits the *trans*-mediated cell-adhesion between perinexal β1-subunits might reduce the signal propagation through cardiac muscle, whilst not completely preventing it. This could potentially reduce triggering of post-infarct arrythmias [125]. Conversely, drugs that stabilise these interactions could be useful as a treatment for other forms of arrythmias such as Brugada syndrome in which re-entrant arrhythmia results from a conduction slowing substrate [66]. It might also be possible to target specific β-subunit signalling pathways, for example the phosphorylation of the β1-subunit cytoplasmic region [126]. These are quite speculative, yet potentially attractive hypotheses that require further investigations. More broadly, the increasing emphasis on the cell-adhesion roles of Nav β-subunits in both healthy and pathological states, offers a more balanced perspective on these proteins and could open completely new avenues for therapy.

Author Contributions: Conceptualization, investigation and writing, S.C.S., C.L.-H.H. and A.P.J. All authors have read and agreed to the published version of the manuscript.

Funding: S.C.S was supported by British Heart Foundation project grants PG/14/79/31102 and PG/19/59/34582 (to S.C.S., C.L.-H.H., and A.P.J.) and by Isaac Newton Trust Grant G101770.

Acknowledgments: The figures were created with BioRender.com.

References

1. Sweeney, H.L.; Hammers, D.W. Muscle Contraction. *Cold Spring Harb. Perspect. Biol.* **2018**, *10*. [CrossRef] [PubMed]
2. Cortada, E.; Brugada, R.; Verges, M. Trafficking and Function of the Voltage-Gated Sodium Channel beta2 Subunit. *Biomolecules* **2019**, *9*, 604. [CrossRef] [PubMed]
3. DeMarco, K.R.; Clancy, C.E. Cardiac Na Channels: Structure to Function. *Curr. Top. Membr.* **2016**, *78*, 287–311. [CrossRef] [PubMed]
4. Veeraraghavan, R.; Gyorke, S.; Radwanski, P.B. Neuronal sodium channels: Emerging components of the nano-machinery of cardiac calcium cycling. *J. Physiol.* **2017**, *595*, 3823–3834. [CrossRef]
5. Maier, S.K.; Westenbroek, R.E.; McCormick, K.A.; Curtis, R.; Scheuer, T.; Catterall, W.A. Distinct subcellular localization of different sodium channel alpha and beta subunits in single ventricular myocytes from mouse heart. *Circulation* **2004**, *109*, 1421–1427. [CrossRef] [PubMed]
6. Fraser, J.A.; Huang, C.L.; Pedersen, T.H. Relationships between resting conductances, excitability, and t-system ionic homeostasis in skeletal muscle. *J. Gen. Physiol.* **2011**, *138*, 95–116. [CrossRef] [PubMed]
7. Pedersen, T.H.; Huang, C.L.-H.; Fraser, J.A. An analysis of the relationships between subthreshold electrical properties and excitability in skeletal muscle. *J. Gen. Physiol.* **2011**, *138*, 73–93. [CrossRef]
8. Ahern, C.A.; Payandeh, J.; Bosmans, F.; Chanda, B. The hitchhiker's guide to the voltage-gated sodium channel galaxy. *J. Gen. Physiol.* **2016**, *147*, 1–24. [CrossRef]
9. Nishino, A.; Okamura, Y. Evolutionary History of Voltage-Gated Sodium Channels. *Handb. Exp. Pharmacol.* **2018**, *246*, 3–32. [CrossRef]

10. Yan, Z.; Zhou, Q.; Wang, L.; Wu, J.; Zhao, Y.; Huang, G.; Peng, W.; Shen, H.; Lei, J.; Yan, N. Structure of the Nav1.4-beta1 Complex from Electric Eel. *Cell* **2017**, *170*, 470–482.e11. [CrossRef]

11. Pan, X.; Li, Z.; Zhou, Q.; Shen, H.; Wu, K.; Huang, X.; Chen, J.; Zhang, J.; Zhu, X.; Lei, J.; et al. Structure of the human voltage-gated sodium channel Nav1.4 in complex with beta1. *Science* **2018**, *362*. [CrossRef] [PubMed]

12. Shen, H.; Liu, D.; Wu, K.; Lei, J.; Yan, N. Structures of human Nav1.7 channel in complex with auxiliary subunits and animal toxins. *Science* **2019**, *363*, 1303–1308. [CrossRef] [PubMed]

13. Pan, X.; Li, Z.; Huang, X.; Huang, G.; Gao, S.; Shen, H.; Liu, L.; Lei, J.; Yan, N. Molecular basis for pore blockade of human Na(+) channel Nav1.2 by the mu-conotoxin KIIIA. *Science* **2019**, *363*, 1309–1313. [CrossRef] [PubMed]

14. Jiang, D.; Shi, H.; Tonggu, L.; Gamal El-Din, T.M.; Lenaeus, M.J.; Zhao, Y.; Yoshioka, C.; Zheng, N.; Catterall, W.A. Structure of the Cardiac Sodium Channel. *Cell* **2020**, *180*, 122–134.e110. [CrossRef]

15. Edokobi, N.; Isom, L.L. Voltage-Gated Sodium Channel beta1/beta1B Subunits Regulate Cardiac Physiology and Pathophysiology. *Front. Physiol.* **2018**, *9*, 351. [CrossRef]

16. Brackenbury, W.J.; Isom, L.L. Na Channel beta Subunits: Overachievers of the Ion Channel Family. *Front. Pharmacol.* **2011**, *2*, 53. [CrossRef]

17. Namadurai, S.; Yereddi, N.R.; Cusdin, F.S.; Huang, C.L.; Chirgadze, D.Y.; Jackson, A.P. A new look at sodium channel beta subunits. *Open Biol.* **2015**, *5*, 140192. [CrossRef]

18. Patino, G.A.; Isom, L.L. Electrophysiology and beyond: Multiple roles of Na+ channel beta subunits in development and disease. *Neurosci. Lett.* **2010**, *486*, 53–59. [CrossRef]

19. Salvage, S.C.; Zhu, W.; Habib, Z.F.; Hwang, S.S.; Irons, J.R.; Huang, C.L.H.; Silva, J.R.; Jackson, A.P. Gating control of the cardiac sodium channel Nav1.5 by its beta3-subunit involves distinct roles for a transmembrane glutamic acid and the extracellular domain. *J. Biol. Chem.* **2019**. [CrossRef]

20. Salvage, S.C.; Rees, J.S.; McStea, A.; Hirsch, M.; Wang, L.; Tynan, C.J.; Reed, M.W.; Irons, J.R.; Butler, R.; Thompson, A.J.; et al. Supramolecular clustering of the cardiac sodium channel Nav1.5 in HEK293F cells, with and without the auxiliary beta3-subunit. *FASEB J.* **2020**, *34*, 3537–3553. [CrossRef]

21. Zhu, W.; Voelker, T.L.; Varga, Z.; Schubert, A.R.; Nerbonne, J.M.; Silva, J.R. Mechanisms of noncovalent beta subunit regulation of NaV channel gating. *J. Gen. Physiol.* **2017**. [CrossRef] [PubMed]

22. Hakim, P.; Gurung, I.S.; Pedersen, T.H.; Thresher, R.; Brice, N.; Lawrence, J.; Grace, A.A.; Huang, C.L. Scn3b knockout mice exhibit abnormal ventricular electrophysiological properties. *Prog. Biophys. Mol. Biol.* **2008**, *98*, 251–266. [CrossRef] [PubMed]

23. Ko, S.H.; Lenkowski, P.W.; Lee, H.C.; Mounsey, J.P.; Patel, M.K. Modulation of Na(v)1.5 by beta1- and beta3-subunit co-expression in mammalian cells. *Pflug. Arch.* **2005**, *449*, 403–412. [CrossRef] [PubMed]

24. Chen, C.; Calhoun, J.D.; Zhang, Y.; Lopez-Santiago, L.; Zhou, N.; Davis, T.H.; Salzer, J.L.; Isom, L.L. Identification of the cysteine residue responsible for disulfide linkage of Na+ channel alpha and beta2 subunits. *J. Biol. Chem.* **2012**, *287*, 39061–39069. [CrossRef] [PubMed]

25. Yu, F.H.; Westenbroek, R.E.; Silos-Santiago, I.; McCormick, K.A.; Lawson, D.; Ge, P.; Ferriera, H.; Lilly, J.; DiStefano, P.S.; Catterall, W.A.; et al. Sodium channel beta4, a new disulfide-linked auxiliary subunit with similarity to beta2. *J. Neurosci.* **2003**, *23*, 7577–7585. [CrossRef]

26. Gilchrist, J.; Das, S.; Van Petegem, F.; Bosmans, F. Crystallographic insights into sodium-channel modulation by the beta4 subunit. *Proc. Natl. Acad. Sci. USA* **2013**, *110*, E5016–E5024. [CrossRef]

27. Das, S.; Gilchrist, J.; Bosmans, F.; Van Petegem, F. Binary architecture of the Nav1.2-beta2 signaling complex. *eLife* **2016**, *5*. [CrossRef]

28. Molinarolo, S.; Granata, D.; Carnevale, V.; Ahern, C.A. Mining Protein Evolution for Insights into Mechanisms of Voltage-Dependent Sodium Channel Auxiliary Subunits. *Handb. Exp. Pharmacol.* **2018**, *246*, 33–49. [CrossRef]

29. Kusano, K.; Thomas, T.N.; Fujiwara, K. Phosphorylation and localization of protein-zero related (PZR) in cultured endothelial cells. *Endothelium* **2008**, *15*, 127–136. [CrossRef]

30. Chopra, S.S.; Watanabe, H.; Zhong, T.P.; Roden, D.M. Molecular cloning and analysis of zebrafish voltage-gated sodium channel beta subunit genes: Implications for the evolution of electrical signaling in vertebrates. *BMC Evol. Biol.* **2007**, *7*, 113. [CrossRef]

31. McEwen, D.P.; Isom, L.L. Heterophilic interactions of sodium channel beta1 subunits with axonal and glial cell adhesion molecules. *J. Biol. Chem.* **2004**, *279*, 52744–52752. [CrossRef] [PubMed]

32. Jansson, K.H.; Castillo, D.G.; Morris, J.W.; Boggs, M.E.; Czymmek, K.J.; Adams, E.L.; Schramm, L.P.; Sikes, R.A. Identification of beta-2 as a key cell adhesion molecule in PCa cell neurotropic behavior: A novel ex vivo and biophysical approach. *PLoS ONE* **2014**, *9*, e98408. [CrossRef] [PubMed]

33. Namadurai, S.; Balasuriya, D.; Rajappa, R.; Wiemhofer, M.; Stott, K.; Klingauf, J.; Edwardson, J.M.; Chirgadze, D.Y.; Jackson, A.P. Crystal structure and molecular imaging of the Nav channel beta3 subunit indicates a trimeric assembly. *J. Biol. Chem.* **2014**, *289*, 10797–10811. [CrossRef] [PubMed]

34. Ratcliffe, C.F.; Westenbroek, R.E.; Curtis, R.; Catterall, W.A. Sodium channel beta1 and beta3 subunits associate with neurofascin through their extracellular immunoglobulin-like domain. *J. Cell Biol.* **2001**, *154*, 427–434. [CrossRef]

35. Shimizu, H.; Miyazaki, H.; Ohsawa, N.; Shoji, S.; Ishizuka-Katsura, Y.; Tosaki, A.; Oyama, F.; Terada, T.; Sakamoto, K.; Shirouzu, M.; et al. Structure-based site-directed photo-crosslinking analyses of multimeric cell-adhesive interactions of voltage-gated sodium channel beta subunits. *Sci. Rep.* **2016**, *6*, 26618. [CrossRef]

36. Isom, L.L. The role of sodium channels in cell adhesion. *Front. Biosci.* **2002**, *7*, 12–23. [CrossRef]

37. Manring, H.R.; Dorn, L.E.; Ex-Willey, A.; Accornero, F.; Ackermann, M.A. At the heart of inter- and intracellular signaling: The intercalated disc. *Biophys. Rev.* **2018**, *10*, 961–971. [CrossRef]

38. Li, Y.; Merkel, C.D.; Zeng, X.; Heier, J.A.; Cantrell, P.S.; Sun, M.; Stolz, D.B.; Watkins, S.C.; Yates, N.A.; Kwiatkowski, A.V. The N-cadherin interactome in primary cardiomyocytes as defined using quantitative proximity proteomics. *J. Cell Sci.* **2019**, *132*. [CrossRef]

39. Schinner, C.; Erber, B.M.; Yeruva, S.; Waschke, J. Regulation of cardiac myocyte cohesion and gap junctions via desmosomal adhesion. *Acta Physiol.* **2019**, *226*, e13242. [CrossRef]

40. Kleber, A.G.; Saffitz, J.E. Role of the intercalated disc in cardiac propagation and arrhythmogenesis. *Front. Physiol.* **2014**, *5*, 404. [CrossRef]

41. Vozzi, C.; Dupont, E.; Coppen, S.R.; Yeh, H.I.; Severs, N.J. Chamber-related differences in connexin expression in the human heart. *J. Mol. Cell Cardiol.* **1999**, *31*, 991–1003. [CrossRef] [PubMed]

42. Goodenough, D.A.; Goliger, J.A.; Paul, D.L. Connexins, connexons, and intercellular communication. *Annu. Rev. Biochem.* **1996**, *65*, 475–502. [CrossRef] [PubMed]

43. Goldberg, G.S.; Valiunas, V.; Brink, P.R. Selective permeability of gap junction channels. *Biochim. Biophys. Acta* **2004**, *1662*, 96–101. [CrossRef]

44. Zhang, Q.; Bai, X.; Liu, Y.; Wang, K.; Shen, B.; Sun, X. Current Concepts and Perspectives on Connexin43: A Mini Review. *Curr. Protein Pept. Sci.* **2018**, *19*, 1049–1057. [CrossRef]

45. Kleber, A.G.; Rudy, Y. Basic mechanisms of cardiac impulse propagation and associated arrhythmias. *Physiol. Rev.* **2004**, *84*, 431–488. [CrossRef] [PubMed]

46. Rohr, S. Role of gap junctions in the propagation of the cardiac action potential. *Cardiovasc. Res.* **2004**, *62*, 309–322. [CrossRef]

47. Rhett, J.M.; Veeraraghavan, R.; Poelzing, S.; Gourdie, R.G. The perinexus: Sign-post on the path to a new model of cardiac conduction? *Trends Cardiovasc. Med.* **2013**, *23*, 222–228. [CrossRef]

48. Rhett, J.M.; Gourdie, R.G. The perinexus: A new feature of Cx43 gap junction organization. *Heart Rhythm.* **2012**, *9*, 619–623. [CrossRef]

49. Rhett, J.M.; Ongstad, E.L.; Jourdan, J.; Gourdie, R.G. Cx43 associates with Na(v)1.5 in the cardiomyocyte perinexus. *J. Membr. Biol.* **2012**, *245*, 411–422. [CrossRef]

50. Hunter, A.W.; Barker, R.J.; Zhu, C.; Gourdie, R.G. Zonula occludens-1 alters connexin43 gap junction size and organization by influencing channel accretion. *Mol. Biol. Cell* **2005**, *16*, 5686–5698. [CrossRef]

51. Meadows, L.S.; Isom, L.L. Sodium channels as macromolecular complexes: Implications for inherited arrhythmia syndromes. *Cardiovasc. Res.* **2005**, *67*, 448–458. [CrossRef] [PubMed]

52. Sorgen, P.L.; Trease, A.J.; Spagnol, G.; Delmar, M.; Nielsen, M.S. Protein(-)Protein Interactions with Connexin 43: Regulation and Function. *Int. J. Mol. Sci.* **2018**, *19*, 1428. [CrossRef] [PubMed]

53. Makara, M.A.; Curran, J.; Little, S.C.; Musa, H.; Polina, I.; Smith, S.A.; Wright, P.J.; Unudurthi, S.D.; Snyder, J.; Bennett, V.; et al. Ankyrin-G coordinates intercalated disc signaling platform to regulate cardiac excitability In Vivo. *Circ. Res.* **2014**, *115*, 929–938. [CrossRef] [PubMed]

54. Clatot, J.; Hoshi, M.; Wan, X.; Liu, H.; Jain, A.; Shinlapawittayatorn, K.; Marionneau, C.; Ficker, E.; Ha, T.; Deschenes, I. Voltage-gated sodium channels assemble and gate as dimers. *Nat. Commun.* **2017**, *8*, 2077. [CrossRef]

55. Mori, Y.; Fishman, G.I.; Peskin, C.S. Ephaptic conduction in a cardiac strand model with 3D electrodiffusion. *Proc. Natl. Acad. Sci. USA* **2008**, *105*, 6463–6468. [CrossRef]

56. Lin, J.; Keener, J.P. Modeling electrical activity of myocardial cells incorporating the effects of ephaptic coupling. *Proc. Natl. Acad. Sci. USA* **2010**, *107*, 20935–20940. [CrossRef]

57. Hichri, E.; Abriel, H.; Kucera, J.P. Distribution of cardiac sodium channels in clusters potentiates ephaptic interactions in the intercalated disc. *J. Physiol.* **2018**, *596*, 563–589. [CrossRef]

58. Sperelakis, N. An electric field mechanism for transmission of excitation between myocardial cells. *Circ. Res.* **2002**, *91*, 985–987. [CrossRef] [PubMed]

59. Veeraraghavan, R.; Lin, J.; Keener, J.P.; Gourdie, R.; Poelzing, S. Potassium channels in the Cx43 gap junction perinexus modulate ephaptic coupling: An experimental and modeling study. *Pflug. Arch.* **2016**, *468*, 1651–1661. [CrossRef]

60. Willis, B.C.; Ponce-Balbuena, D.; Jalife, J. Protein assemblies of sodium and inward rectifier potassium channels control cardiac excitability and arrhythmogenesis. *Am. J. Physiol. Heart Circ. Physiol.* **2015**, *308*, H1463–H1473. [CrossRef]

61. Abriel, H.; Rougier, J.S.; Jalife, J. Ion channel macromolecular complexes in cardiomyocytes: Roles in sudden cardiac death. *Circ. Res.* **2015**, *116*, 1971–1988. [CrossRef] [PubMed]

62. Milstein, M.L.; Musa, H.; Balbuena, D.P.; Anumonwo, J.M.; Auerbach, D.S.; Furspan, P.B.; Hou, L.; Hu, B.; Schumacher, S.M.; Vaidyanathan, R.; et al. Dynamic reciprocity of sodium and potassium channel expression in a macromolecular complex controls cardiac excitability and arrhythmia. *Proc. Natl. Acad. Sci. USA* **2012**, *109*, E2134–E2143. [CrossRef] [PubMed]

63. Lopatin, A.N.; Nichols, C.G. Inward rectifiers in the heart: An update on I(K1). *J. Mol. Cell Cardiol.* **2001**, *33*, 625–638. [CrossRef] [PubMed]

64. Burstein, B.; Nattel, S. Atrial fibrosis: Mechanisms and clinical relevance in atrial fibrillation. *J. Am. Coll. Cardiol.* **2008**, *51*, 802–809. [CrossRef] [PubMed]

65. Jeevaratnam, K.; Poh Tee, S.; Zhang, Y.; Rewbury, R.; Guzadhur, L.; Duehmke, R.; Grace, A.A.; Lei, M.; Huang, C.L. Delayed conduction and its implications in murine Scn5a(+/−) hearts: Independent and interacting effects of genotype, age, and sex. *Pflug. Arch.* **2011**, *461*, 29–44. [CrossRef]

66. Veeraraghavan, R.; Hoeker, G.S.; Alvarez-Laviada, A.; Hoagland, D.; Wan, X.; King, D.R.; Sanchez-Alonso, J.; Chen, C.; Jourdan, J.; Isom, L.L.; et al. The adhesion function of the sodium channel beta subunit (beta1) contributes to cardiac action potential propagation. *eLife* **2018**, *7*. [CrossRef]

67. Malhotra, J.D.; Kazen-Gillespie, K.; Hortsch, M.; Isom, L.L. Sodium channel beta subunits mediate homophilic cell adhesion and recruit ankyrin to points of cell-cell contact. *J. Biol. Chem.* **2000**, *275*, 11383–11388. [CrossRef]

68. Raisch, T.B.; Yanoff, M.S.; Larsen, T.R.; Farooqui, M.A.; King, D.R.; Veeraraghavan, R.; Gourdie, R.G.; Baker, J.W.; Arnold, W.S.; AlMahameed, S.T.; et al. Intercalated Disk Extracellular Nanodomain Expansion in Patients With Atrial Fibrillation. *Front. Physiol.* **2018**, *9*, 398. [CrossRef]

69. Watanabe, H.; Koopmann, T.T.; Le Scouarnec, S.; Yang, T.; Ingram, C.R.; Schott, J.J.; Demolombe, S.; Probst, V.; Anselme, F.; Escande, D.; et al. Sodium channel beta1 subunit mutations associated with Brugada syndrome and cardiac conduction disease in humans. *J. Clin. Investig.* **2008**, *118*, 2260–2268. [CrossRef]

70. Watanabe, H.; Darbar, D.; Kaiser, D.W.; Jiramongkolchai, K.; Chopra, S.; Donahue, B.S.; Kannankeril, P.J.; Roden, D.M. Mutations in sodium channel beta1- and beta2-subunits associated with atrial fibrillation. *Circ. Arrhythm Electrophysiol.* **2009**, *2*, 268–275. [CrossRef]

71. Kobirumaki-Shimozawa, F.; Nakanishi, T.; Shimozawa, T.; Terui, T.; Oyama, K.; Li, J.; Louch, W.E.; Ishiwata, S.; Fukuda, N. Real-Time In Vivo Imaging of Mouse Left Ventricle Reveals Fluctuating Movements of the Intercalated Discs. *Nanomaterials* **2020**, *10*, 532. [CrossRef] [PubMed]

72. Malhotra, J.D.; Koopmann, M.C.; Kazen-Gillespie, K.A.; Fettman, N.; Hortsch, M.; Isom, L.L. Structural requirements for interaction of sodium channel beta 1 subunits with ankyrin. *J. Biol. Chem.* **2002**, *277*, 26681–26688. [CrossRef] [PubMed]

73. Brackenbury, W.J.; Djamgoz, M.B.; Isom, L.L. An emerging role for voltage-gated Na+ channels in cellular migration: Regulation of central nervous system development and potentiation of invasive cancers. *Neuroscientist* **2008**, *14*, 571–583. [CrossRef] [PubMed]

74. Malhotra, J.D.; Thyagarajan, V.; Chen, C.; Isom, L.L. Tyrosine-phosphorylated and nonphosphorylated sodium channel beta1 subunits are differentially localized in cardiac myocytes. *J. Biol. Chem.* **2004**, *279*, 40748–40754. [CrossRef]

75. Hu, D.; Barajas-Martinez, H.; Medeiros-Domingo, A.; Crotti, L.; Veltmann, C.; Schimpf, R.; Urrutia, J.; Alday, A.; Casis, O.; Pfeiffer, R.; et al. A novel rare variant in SCN1Bb linked to Brugada syndrome and SIDS by combined modulation of Na(v)1.5 and K(v)4.3 channel currents. *Heart Rhythm.* **2012**, *9*, 760–769. [CrossRef]

76. Yereddi, N.R.; Cusdin, F.S.; Namadurai, S.; Packman, L.C.; Monie, T.P.; Slavny, P.; Clare, J.J.; Powell, A.J.; Jackson, A.P. The immunoglobulin domain of the sodium channel beta3 subunit contains a surface-localized disulfide bond that is required for homophilic binding. *FASEB J.* **2013**, *27*, 568–580. [CrossRef]

77. McEwen, D.P.; Chen, C.; Meadows, L.S.; Lopez-Santiago, L.; Isom, L.L. The voltage-gated Na+ channel beta3 subunit does not mediate trans homophilic cell adhesion or associate with the cell adhesion molecule contactin. *Neurosci. Lett.* **2009**, *462*, 272–275. [CrossRef]

78. Dhar Malhotra, J.; Chen, C.; Rivolta, I.; Abriel, H.; Malhotra, R.; Mattei, L.N.; Brosius, F.C.; Kass, R.S.; Isom, L.L. Characterization of sodium channel alpha- and beta-subunits in rat and mouse cardiac myocytes. *Circulation* **2001**, *103*, 1303–1310. [CrossRef]

79. Shimizu, H.; Tosaki, A.; Ohsawa, N.; Ishizuka-Katsura, Y.; Shoji, S.; Miyazaki, H.; Oyama, F.; Terada, T.; Shirouzu, M.; Sekine, S.I.; et al. Parallel homodimer structures of the extracellular domains of the voltage-gated sodium channel beta4 subunit explain its role in cell-cell adhesion. *J. Biol. Chem.* **2017**, *292*, 13428–13440. [CrossRef]

80. Shapiro, L.; Fannon, A.M.; Kwong, P.D.; Thompson, A.; Lehmann, M.S.; Grubel, G.; Legrand, J.F.; Als-Nielsen, J.; Colman, D.R.; Hendrickson, W.A. Structural basis of cell-cell adhesion by cadherins. *Nature* **1995**, *374*, 327–337. [CrossRef]

81. Harrison, O.J.; Brasch, J.; Lasso, G.; Katsamba, P.S.; Ahlsen, G.; Honig, B.; Shapiro, L. Structural basis of adhesive binding by desmocollins and desmogleins. *Proc. Natl. Acad. Sci. USA* **2016**, *113*, 7160–7165. [CrossRef] [PubMed]

82. Brasch, J.; Harrison, O.J.; Honig, B.; Shapiro, L. Thinking outside the cell: How cadherins drive adhesion. *Trends Cell Biol.* **2012**, *22*, 299–310. [CrossRef]

83. Medeiros-Domingo, A.; Kaku, T.; Tester, D.J.; Iturralde-Torres, P.; Itty, A.; Ye, B.; Valdivia, C.; Ueda, K.; Canizales-Quinteros, S.; Tusie-Luna, M.T.; et al. SCN4B-encoded sodium channel beta4 subunit in congenital long-QT syndrome. *Circulation* **2007**, *116*, 134–142. [CrossRef] [PubMed]

84. Yarbrough, T.L.; Lu, T.; Lee, H.C.; Shibata, E.F. Localization of cardiac sodium channels in caveolin-rich membrane domains: Regulation of sodium current amplitude. *Circ. Res.* **2002**, *90*, 443–449. [CrossRef] [PubMed]

85. Maguy, A.; Hebert, T.E.; Nattel, S. Involvement of lipid rafts and caveolae in cardiac ion channel function. *Cardiovasc. Res.* **2006**, *69*, 798–807. [CrossRef] [PubMed]

86. Sanon, V.P.; Sawaki, D.; Mjaatvedt, C.H.; Jourdan-Le Saux, C. Myocardial tissue caveolae. *Compr. Physiol.* **2015**, *5*, 871–886. [CrossRef]

87. Aicart-Ramos, C.; Valero, R.A.; Rodriguez-Crespo, I. Protein palmitoylation and subcellular trafficking. *Biochim. Biophys. Acta* **2011**, *1808*, 2981–2994. [CrossRef]

88. Bouza, A.A.; Philippe, J.M.; Edokobi, N.; Pinsky, A.M.; Offord, J.; Calhoun, J.D.; Lopez-Floran, M.; Lopez-Santiago, L.F.; Jenkins, P.M.; Isom, L.L. Sodium channel beta1 subunits are post-translationally modified by tyrosine phosphorylation, S-palmitoylation, and regulated intramembrane proteolysis. *J. Biol. Chem.* **2020**. [CrossRef]

89. Alday, A.; Urrutia, J.; Gallego, M.; Casis, O. alpha1-adrenoceptors regulate only the caveolae-located subpopulation of cardiac K(V)4 channels. *Channels (Austin)* **2010**, *4*, 168–178. [CrossRef]

90. Vaidyanathan, R.; Reilly, L.; Eckhardt, L.L. Caveolin-3 Microdomain: Arrhythmia Implications for Potassium Inward Rectifier and Cardiac Sodium Channel. *Front. Physiol.* **2018**, *9*, 1548. [CrossRef]

91. Marionneau, C.; Carrasquillo, Y.; Norris, A.J.; Townsend, R.R.; Isom, L.L.; Link, A.J.; Nerbonne, J.M. The sodium channel accessory subunit Navbeta1 regulates neuronal excitability through modulation of repolarizing voltage-gated K(+) channels. *J. Neurosci.* **2012**, *32*, 5716–5727. [CrossRef] [PubMed]

92. Dabiri, B.E.; Lee, H.; Parker, K.K. A potential role for integrin signaling in mechanoelectrical feedback. *Prog. Biophys. Mol. Biol.* **2012**, *110*, 196–203. [CrossRef] [PubMed]

93. Kawabe, J.; Okumura, S.; Lee, M.C.; Sadoshima, J.; Ishikawa, Y. Translocation of caveolin regulates stretch-induced ERK activity in vascular smooth muscle cells. *Am. J. Physiol. Heart Circ. Physiol.* **2004**, *286*, H1845–H1852. [CrossRef]

94. Israeli-Rosenberg, S.; Chen, C.; Li, R.; Deussen, D.N.; Niesman, I.R.; Okada, H.; Patel, H.H.; Roth, D.M.; Ross, R.S. Caveolin modulates integrin function and mechanical activation in the cardiomyocyte. *FASEB J.* **2015**, *29*, 374–384. [CrossRef] [PubMed]

95. Davis, M.J.; Wu, X.; Nurkiewicz, T.R.; Kawasaki, J.; Gui, P.; Hill, M.A.; Wilson, E. Regulation of ion channels by integrins. *Cell Biochem. Biophys.* **2002**, *36*, 41–66. [CrossRef]

96. Tyan, L.; Foell, J.D.; Vincent, K.P.; Woon, M.T.; Mesquitta, W.T.; Lang, D.; Best, J.M.; Ackerman, M.J.; McCulloch, A.D.; Glukhov, A.V.; et al. Long QT syndrome caveolin-3 mutations differentially modulate Kv 4 and Cav 1.2 channels to contribute to action potential prolongation. *J. Physiol.* **2019**, *597*, 1531–1551. [CrossRef] [PubMed]

97. Vatta, M.; Ackerman, M.J.; Ye, B.; Makielski, J.C.; Ughanze, E.E.; Taylor, E.W.; Tester, D.J.; Balijepalli, R.C.; Foell, J.D.; Li, Z.; et al. Mutant caveolin-3 induces persistent late sodium current and is associated with long-QT syndrome. *Circulation* **2006**, *114*, 2104–2112. [CrossRef]

98. Echarri, A.; Del Pozo, M.A. Caveolae—mechanosensitive membrane invaginations linked to actin filaments. *J. Cell Sci.* **2015**, *128*, 2747–2758. [CrossRef]

99. Kohl, P.; Cooper, P.J.; Holloway, H. Effects of acute ventricular volume manipulation on In Situ cardiomyocyte cell membrane configuration. *Prog. Biophys. Mol. Biol.* **2003**, *82*, 221–227. [CrossRef]

100. Sinha, B.; Koster, D.; Ruez, R.; Gonnord, P.; Bastiani, M.; Abankwa, D.; Stan, R.V.; Butler-Browne, G.; Vedie, B.; Johannes, L.; et al. Cells respond to mechanical stress by rapid disassembly of caveolae. *Cell* **2011**, *144*, 402–413. [CrossRef]

101. Wary, K.K.; Mariotti, A.; Zurzolo, C.; Giancotti, F.G. A requirement for caveolin-1 and associated kinase Fyn in integrin signaling and anchorage-dependent cell growth. *Cell* **1998**, *94*, 625–634. [CrossRef]

102. Ahern, C.A.; Zhang, J.F.; Wookalis, M.J.; Horn, R. Modulation of the cardiac sodium channel NaV1.5 by Fyn, a Src family tyrosine kinase. *Circ. Res.* **2005**, *96*, 991–998. [CrossRef]

103. Beyder, A.; Rae, J.L.; Bernard, C.; Strege, P.R.; Sachs, F.; Farrugia, G. Mechanosensitivity of Nav1.5, a voltage-sensitive sodium channel. *J. Physiol.* **2010**, *588*, 4969–4985. [CrossRef] [PubMed]

104. Morris, C.E.; Juranka, P.F. Nav channel mechanosensitivity: Activation and inactivation accelerate reversibly with stretch. *Biophys. J.* **2007**, *93*, 822–833. [CrossRef]

105. Maroni, M.; Korner, J.; Schuttler, J.; Winner, B.; Lampert, A.; Eberhardt, E. beta1 and beta3 subunits amplify mechanosensitivity of the cardiac voltage-gated sodium channel Nav1.5. *Pflug. Arch.* **2019**, *471*, 1481–1492. [CrossRef] [PubMed]

106. Koivumaki, J.T.; Clark, R.B.; Belke, D.; Kondo, C.; Fedak, P.W.; Maleckar, M.M.; Giles, W.R. Na(+) current expression in human atrial myofibroblasts: Identity and functional roles. *Front. Physiol.* **2014**, *5*, 275. [CrossRef] [PubMed]

107. Nattel, S. Electrical coupling between cardiomyocytes and fibroblasts: Experimental testing of a challenging and important concept. *Cardiovasc. Res.* **2018**, *114*, 349–352. [CrossRef] [PubMed]

108. Rog-Zielinska, E.A.; Kong, C.H.T.; Zgierski-Johnston, C.M.; Verkade, P.; Mantell, J.; Cannell, M.B.; Kohl, P. Species differences in the morphology of transverse tubule openings in cardiomyocytes. *Europace* **2018**, *20*, iii120–iii124. [CrossRef]

109. Kostin, S.; Scholz, D.; Shimada, T.; Maeno, Y.; Mollnau, H.; Hein, S.; Schaper, J. The internal and external protein scaffold of the T-tubular system in cardiomyocytes. *Cell Tissue Res.* **1998**, *294*, 449–460. [CrossRef]

110. Lin, X.; O'Malley, H.; Chen, C.; Auerbach, D.; Foster, M.; Shekhar, A.; Zhang, M.; Coetzee, W.; Jalife, J.; Fishman, G.I.; et al. Scn1b deletion leads to increased tetrodotoxin-sensitive sodium current, altered intracellular calcium homeostasis and arrhythmias in murine hearts. *J. Physiol.* **2015**, *593*, 1389–1407. [CrossRef]

111. Abriel, H. Cardiac sodium channel Na(v)1.5 and interacting proteins: Physiology and pathophysiology. *J. Mol. Cell Cardiol.* **2010**, *48*, 2–11. [CrossRef] [PubMed]

112. Kitazawa, M.; Kubo, Y.; Nakajo, K. The stoichiometry and biophysical properties of the Kv4 potassium channel complex with K+ channel-interacting protein (KChIP) subunits are variable, depending on the relative expression level. *J. Biol. Chem.* **2014**, *289*, 17597–17609. [CrossRef] [PubMed]

113. Doolittle, R.F. The multiplicity of domains in proteins. *Annu. Rev. Biochem.* **1995**, *64*, 287–314. [CrossRef] [PubMed]

114. Zhao, Z.J.; Zhao, R. Purification and cloning of PZR, a binding protein and putative physiological substrate of tyrosine phosphatase SHP-2. *J. Biol. Chem.* **1998**, *273*, 29367–29372. [CrossRef]

115. Wang, L.; Nomura, Y.; Du, Y.; Dong, K. Differential effects of TipE and a TipE-homologous protein on modulation of gating properties of sodium channels from Drosophila melanogaster. *PLoS ONE* **2013**, *8*, e67551. [CrossRef]

116. Muhamed, I.; Chowdhury, F.; Maruthamuthu, V. Biophysical Tools to Study Cellular Mechanotransduction. *Bioengineering* **2017**, *4*, 12. [CrossRef]

117. Ohtsubo, K.; Marth, J.D. Glycosylation in cellular mechanisms of health and disease. *Cell* **2006**, *126*, 855–867. [CrossRef]

118. Cortada, E.; Brugada, R.; Verges, M. N-Glycosylation of the voltage-gated sodium channel beta2 subunit is required for efficient trafficking of NaV1.5/beta2 to the plasma membrane. *J. Biol. Chem.* **2019**, *294*, 16123–16140. [CrossRef]

119. Johnson, D.; Montpetit, M.L.; Stocker, P.J.; Bennett, E.S. The sialic acid component of the beta1 subunit modulates voltage-gated sodium channel function. *J. Biol. Chem.* **2004**, *279*, 44303–44310. [CrossRef]

120. Veeraraghavan, R.; Gourdie, R.G.; Poelzing, S. Mechanisms of cardiac conduction: A history of revisions. *Am. J. Physiol. Heart Circ. Physiol.* **2014**, *306*, H619–H627. [CrossRef]

121. Scheffer, I.E.; Harkin, L.A.; Grinton, B.E.; Dibbens, L.M.; Turner, S.J.; Zielinski, M.A.; Xu, R.; Jackson, G.; Adams, J.; Connellan, M.; et al. Temporal lobe epilepsy and GEFS+ phenotypes associated with SCN1B mutations. *Brain* **2007**, *130*, 100–109. [CrossRef] [PubMed]

122. Wang, P.; Wang, S.C.; Li, D.; Li, T.; Yang, H.P.; Wang, L.; Wang, Y.F.; Parpura, V. Role of Connexin 36 in Autoregulation of Oxytocin Neuronal Activity in Rat Supraoptic Nucleus. *ASN Neuro.* **2019**, *11*, 1759091419843762. [CrossRef] [PubMed]

123. Micevych, P.E.; Popper, P.; Hatton, G.I. Connexin 32 mRNA levels in the rat supraoptic nucleus: Up-regulation prior to parturition and during lactation. *Neuroendocrinology* **1996**, *63*, 39–45. [CrossRef] [PubMed]

124. Furshpan, E.J.; Furukawa, T. Intracellular and extracellular responses of the several regions of the Mauthner cell of the goldfish. *J. Neurophysiol.* **1962**, *25*, 732–771. [CrossRef]

125. Ter Keurs, H.E.; Zhang, Y.M.; Davidoff, A.W.; Boyden, P.A.; Wakayama, Y.; Miura, M. Damage induced arrhythmias: Mechanisms and implications. *Can. J. Physiol. Pharmacol.* **2001**, *79*, 73–81. [CrossRef]

126. Brackenbury, W.J.; Isom, L.L. Voltage-gated Na+ channels: Potential for beta subunits as therapeutic targets. *Expert Opin. Ther. Targets* **2008**, *12*, 1191–1203. [CrossRef]

 *biomolecules*

Review

Long QT Syndrome Type 2: Emerging Strategies for Correcting Class 2 *KCNH2* (*hERG*) Mutations and Identifying New Patients

Makoto Ono [1], Don E. Burgess [1], Elizabeth A. Schroder [1], Claude S. Elayi [2], Corey L. Anderson [3], Craig T. January [3], Bin Sun [4], Kalyan Immadisetty [4], Peter M. Kekenes-Huskey [4] and Brian P. Delisle [1,*]

[1] Department of Physiology, Cardiovascular Research Center, Center for Muscle Biology, University of Kentucky, Lexington, KY 40536, USA; makoto.ono@uky.edu (M.O.); deburgess@uky.edu (D.E.B.); eschr0@uky.edu (E.A.S.)

[2] CHI Saint Joseph Hospital, Lexington, KY 40504, USA; samyelayi@sjhlex.org

[3] Cellular and Molecular Arrhythmia Research Program, University of Wisconsin, Madison, WI 53706, USA; clanders@medicine.wisc.edu (C.L.A.); ctj@medicine.wisc.edu (C.T.J.)

[4] Department of Cellular & Molecular Physiology, Loyola University Chicago, Chicago, IL 60153, USA; bsun@luc.edu (B.S.); kimmadisetty@luc.edu (K.I.); pkekeneshuskey@luc.edu (P.M.K.-H.)

* Correspondence: brian.delisle@uky.edu

Received: 17 June 2020; Accepted: 27 July 2020; Published: 4 August 2020

Abstract: Significant advances in our understanding of the molecular mechanisms that cause congenital long QT syndrome (LQTS) have been made. A wide variety of experimental approaches, including heterologous expression of mutant ion channel proteins and the use of inducible pluripotent stem cell-derived cardiomyocytes (iPSC-CMs) from LQTS patients offer insights into etiology and new therapeutic strategies. This review briefly discusses the major molecular mechanisms underlying LQTS type 2 (LQT2), which is caused by loss-of-function (LOF) mutations in the *KCNH2* gene (also known as the human ether-à-go-go-related gene or *hERG*). Almost half of suspected LQT2-causing mutations are missense mutations, and functional studies suggest that about 90% of these mutations disrupt the intracellular transport, or trafficking, of the *KCNH2*-encoded Kv11.1 channel protein to the cell surface membrane. In this review, we discuss emerging strategies that improve the trafficking and functional expression of trafficking-deficient LQT2 Kv11.1 channel proteins to the cell surface membrane and how new insights into the structure of the Kv11.1 channel protein will lead to computational approaches that identify which *KCNH2* missense variants confer a high-risk for LQT2.

Keywords: long QT syndrome; ion channel; trafficking; *KCNH2*; *hERG*

---

1. Introduction

Congenital long QT syndrome (LQTS) is an arrhythmogenic disorder that can manifest as an abnormal prolongation in the heart rate-corrected QT (QTc) interval measured on an electrocardiogram (ECG) [1,2]. It delays ventricular repolarization and increases the probability of life-threatening polymorphic ventricular tachycardia called torsades de pointes (TdP) [3]. LQTS has been identified all over the world with a prevalence as high as 1 in 2500 healthy live births [4]. The major autosomal dominant causes of LQTS are linked to rare mutations in one of three cardiac ion channel genes: *KCNQ1* (LQT1), *KCNH2* (LQT2), and *SCN5A* (LQT3) [5–11]. Most patients with LQTS can be successfully treated with drugs that block β-adrenergic receptors (β-blockers), life-style interventions (e.g., avoid QT prolonging drugs, hypokalemia and hypomagnesemia), and implantable cardioverter defibrillators (ICDs) for high-risk clinical features [12–14].

1.1. KCNH2, Kv11.1 and I_{Kr}

Loss-of-function (LOF) mutations in *KCNH2* are a leading cause of LQTS [15]. *KCNH2* encodes the RNA that is translated into voltage-gated K$^+$ channel Kv11.1 α-subunit channel proteins [5,6]. These α-subunit channel proteins form channels that conduct the rapidly activating delayed rectifier K$^+$ current (I_{Kr}) in the cardiomyocyte sarcolemma membrane. A loss of I_{Kr} can delay the ventricular AP repolarization to cause LQTS (Figure 1). The channels that conduct I_{Kr} in cardiomyocytes are minimally generated by the *KCNH2* gene products Kv11.1a and Kv11.1b α-subunits and the auxiliary subunit MiRP1 (*KCNE2*) [16–19]. Kv11.1a is the full-length, 15-exon, 1159 amino acid α-subunit (*hERG1a* or *KCNH2a*), and Kv11.1b is generated from an alternate start site and 5′ exon that replaces the full-length 1–5 exons (*hERG1b* or *KCNH2b*) [17]. Alternative splicing and polyadenylation of pre-mRNA generate truncated versions of Kv11.1a and Kv11.1b transcripts (Kv11.1a-USO or Kv11.1b-USO), but these gene products are not functional [20,21]. MiRP1 associates with Kv11.1 channel proteins and several other ion channels in the heart [22]. MiRP mutations were previously thought to cause autosomal dominant LQTS, but recent clinical data has invalidated most if not all of these variants [23]. Since the importance of MiRP in LQTS remains unclear, this review focus on the studies that investigate the cellular and molecular mechanisms that cause LOF for LQT2 mutations in Kv11.1a channels.

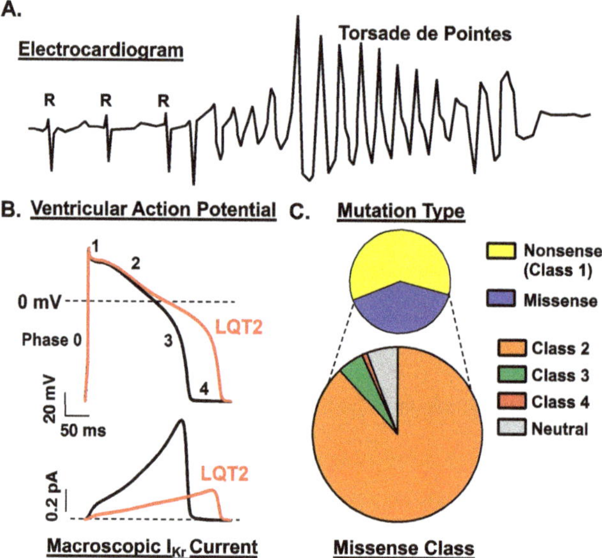

Figure 1. (**A**). This cartoon of an electrocardiogram recording shows normal sinus rhythm with a long QT interval that degenerates into Torsade de Pointes. (**B**). Loss-of-function mutations in *KCNH2* that decrease the amplitude of I_{Kr} increase the ventricular action potential duration. The top panel shows a ventricular action potential with the different phases of the action potential labeled (0, 1, 2, 3, and 4) in control conditions (black traces) or after a loss in I_{Kr} (LQT2, red). The bottom panel shows the corresponding I_{Kr}. The dashed lines represent 0 mV or 0 pA. (**C**). The top pie chart shows the % of candidate LQT2 mutations that are nonsense (Class 1) or missense. To generate this graph, we defined candidate LQT2 missense mutations as ones that are reported to be "likely pathogenic" or "pathogenic" in ClinVar (https://www.ncbi.nlm.nih.gov/clinvar). The bottom pie chart shows the relative percentage of LQT2 missense mutations that are expected to cause LQTS by a Class 2, 3, or 4 mechanism. These data are based on the functional studies of missense variants reported by Anderson and colleagues (2006 and 2014) [24,25]. A few mutations did not have an obvious dysfunctional phenotype (neutral) and might represent *KCNH2* VUS misclassified as causing LQT2.

Kv11.1 channel α-subunits contain cytosolic amino (NH_2) and carboxyl (COOH) termini and six α-helical transmembrane segments (S1–S6) [26,27]. Atom density maps of detergent/lipid-solubilized Kv11.1a channel proteins show that the voltage sensor domain is formed by S1–S4 and the pore domain is formed by S5 and S6 (Figure 2) [27]. The proposed sequence of events for Kv channel biogenesis is: translation and insertion in the endoplasmic reticulum (ER) membrane, the addition of asparagine-linked (N-linked) glycans, the tetramerization of the α-subunits, and the correct or native folding of the voltage-sensor, pore, NH_2 and COOH termini [28].

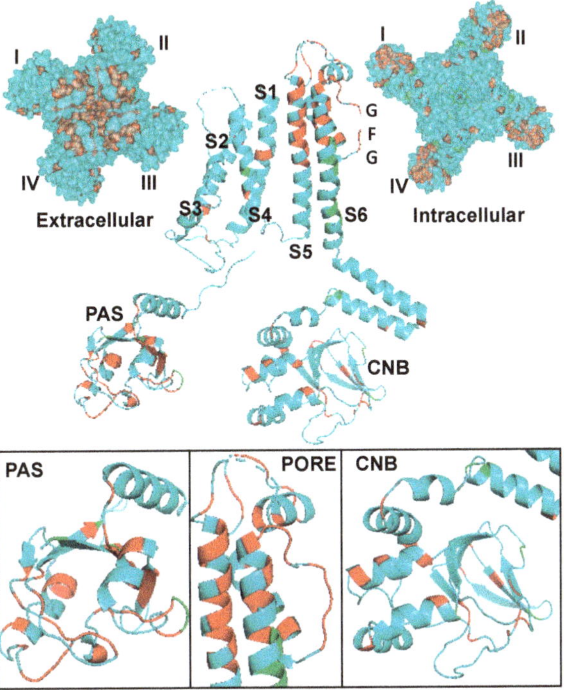

Figure 2. A CryoEM structure of Kv11.1a channel as viewed from the extracellular and intracellular side of the membrane (top left and right, each Kv11.1a α-subunit is denoted I-IV). A side-view of an individual Kv11.1a α-subunit highlighting the selectivity filter (G-F-G) and the secondary structures of the PAS, PORE, and CNB domain (middle and bottom). The location of LQT2 missense mutations that have been tested and disrupt Kv11.1a channel trafficking is shown in red. Green denotes the location of LQT2 missense mutations that have been tested and do not disrupt Kv11.1a trafficking.

The ER extends from the cell nucleus out to the cell periphery [29]. It has several compartments that participate in the biosynthesis of luminal, secretory, membrane associated, and integral membrane proteins and lipids; Ca^{2+} storage and release; apoptotic signaling; quality control; ER-associated degradation; and ER export. The environment in the ER optimizes newly synthesized protein folding and quality control mechanisms prevent the trafficking of incompletely assembled or misfolded proteins to their functional destination in the cell. The folding of newly synthesized proteins and ER associated degradation of misfolded proteins is regulated by molecular chaperone proteins.

Kv11.1a α-subunit channel proteins are synthesized in the ER as an N-linked glycoproteins with a molecular mass of ~135 kDa in the ER (Figure 3) [30]. Chaperone proteins in the ER lumen and the cytosol regulate the folding and ER associated degradation of misfolded Kv11.1 channel proteins [31,32]. Once the Kv11.1 channel proteins achieve their native conformation, they are exported out of the ER into the secretory pathway in coatomer associated protein II (COPII) vesicles to the ER Golgi intermediate compartment (ERGIC) and Golgi compartment [33]. As Kv11.1a proteins traffic through the Golgi apparatus, the N-linked glycans on Kv11.1 α-subunit channel proteins are terminally glycosylated and their molecular mass increases to ~155 kDa [34]. At the cell surface membrane, Kv11.1 channels are continually internalized and recycle back to the cell surface membrane every couple of minutes for several hours before they are targeted for degradation in lysosomes [35,36]. There are several excellent reviews that detail Kv11.1 channel trafficking, chaperones, and the biophysical function of Kv11.1 channels/I_{Kr} [15,16,31,37] This review is focused on the newer strategies being developed to treat LQT2 and identify new LQT2 patients before they suffer a life-threatening event.

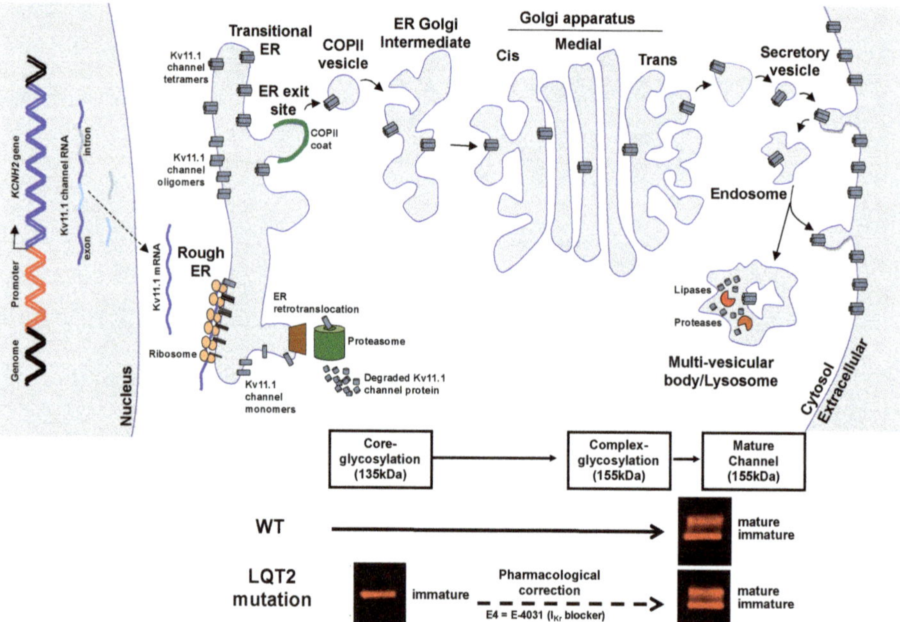

Figure 3. A cartoon diagram of the Kv11.1 channel trafficking pathway. *KCNH2* is transcribed and spliced to mRNA, which is then translated in the ER as a core glycosylated Kv11.1a α-subunit with a molecular mass of 135 kDa. It undergoes terminal glycosylation as it traffics through the Golgi apparatus and its molecular weight increases to 155 kDa. Kv11.1 channels recycle on and off the membrane in endosomes every few minutes for several hours before being degraded in the lysosome pathway. Representative Western blots of cells expressing WT-Kv11.1a α-subunit or a Class 2 LQT2 channel protein are shown below the cartoon. In control conditions, the immunoblot of cells expressing WT-Kv11.1a contains both core and terminally glycosylated Kv11.1a protein bands, whereas cells expressing the Class 2 LQT2 mutant channel protein show only the core glycosylated form. Incubating cells in I_{Kr} blockers (e.g., E-4031) increases terminal glycosylation and functional expression of Class 2 LQT2 channel proteins (pharmacological correction, dashed arrow).

1.2. Long QT Syndrome Type 2

LQT2 mutations are classified by the molecular mechanism with which they decrease I_{Kr} [15]. Class 1 mutations disrupt the synthesis of the *KCNH2*-encoded Kv11.1 channel subunits; Class 2 mutations disrupt the intracellular transport or trafficking of Kv11.1 channel proteins to the cell membrane; Class 3 mutations disrupt I_{Kr} channel gating; and Class 4 mutations disrupt I_{Kr} channel permeation/selectivity. Most LQT2-linked mutations decrease Kv11.1 channel number in the cell surface membrane by a Class 1 or 2 mechanism. A little more than half of LQT2-linked mutations are nonsense and most predict haploinsufficiency via nonsense mediated RNA decay (NMD) (Class 1 mechanism) [38–40]. The remaining LQT2-linked mutations are rare missense mutations, and heterologous expression studies show that about 90% of suspected LQT2 missense mutations disrupt the trafficking of the full length Kv11.1a channel proteins to the cell surface membrane (Class 2 mechanism) (Figure 1) [24,25,41]. These data suggest that the majority of LQT2 cases are linked to mutations that result in a decrease in maximal current. However, the identification of several Class 3 mutations suggests that disruptions in Kv11.1a channel gating also result in LQT2 [42]. Class 3 mutations are expected to reduce the open probability during the repolarization phase of the ventricular AP.

Most Class 2 LQT2 mutations have been identified using Western blot of cells heterologously expressing mutant Kv11.1a channel proteins. Cells expressing Class 2 LQT2 Kv11.1a channel proteins do not generate a 155 kDa Kv11.1a protein band (Figure 3). Some Class 2 Kv11.1a channel proteins generate tetramers that are sequestered throughout the transitional ER compartment, whereas others disrupt the co-assembly of mutant Kv11.1a α-subunits and are targeted for ER associated degradation [32,43–50]. Class 2 Kv11.1a channel proteins that generate tetramers might cause dominant negative (DN) effects by co-assembling and inhibiting the trafficking of WT-Kv11.1a channel proteins [51]. In contrast, Class 2 Kv11.1a mutations that disrupt co-assembly might generate haploinsufficient molecular phenotypes, whereby only the mutant Kv11.1a channel proteins are affected.

1.3. Novel Therapeutic Approaches to Treat LQT2

β-blocker medications and cardiac sympathetic denervation prevent life threatening ventricular arrhythmias, whereas implantable cardioverter defibrillators (ICDs) can terminate these arrhythmias. However, some patients are contraindicated or refractory to β-blocker and ICDs or cardiac sympathetic denervation can result in significant surgical complications [52]. Therefore, researchers are continually working to identify alternative and more effective pharmacological strategies to treat patients with LQT2.

A challenge to treating patients with Class 2 LQT2 mutations is that there is no one dominant disease-causing mutation, but rather hundreds of different Class 2 LQT2 missense mutations that span the Kv11.1a α-subunit (Figure 2) [25]. It is increasingly clear that the identity and location of LQT2-mutations in the Kv11.1a α-subunit are critical determinants for Kv11.1 channel protein trafficking [24]. Most Class 2 LQT2 mutations localize to three major structural domains in the Kv11.1a α-subunit: N-terminal Per-Arnt-Sim domain (PASD), the pore domain, or the C-terminal cyclic nucleotide-binding domain (CNBD). These trends were largely identified through mapping variants identified from genetic screens to the primary sequences [24].

A surprising finding is that incubating cells with drugs that bind to Kv11.1 channel proteins and block I_{Kr} (I_{Kr} blockers) can increase the trafficking for about one third of Class 2 LQT2 mutations (Figure 3) [24,25,53]. Electrophysiological experiments show that pharmacological correction of Class 2 LQT2 Kv11.1a channel protein trafficking also increases functional expression of the mutant channel proteins after drug wash out. A comprehensive analysis of LQT2-linked missense mutations in Kv11.1a channel proteins demonstrated a relationship between their location and ability to undergo pharmacological correction with I_{Kr} blockers [25]. Western blot analysis for over 100 mutant Class 2 LQT2 channel proteins showed that culturing cells in the high-affinity I_{Kr} blocker E-4031 increased the terminal glycosylation for 47% of Kv11.1a channel proteins with Class 2 mutations in the PAS domain, 33% of the Kv11.1 channel proteins with Class 2 mutations in the pore domain, and 21% of the Kv11.1 channel proteins with Class 2 mutations in the CNB domain. Unfortunately, these drugs have limited therapeutic potential since they block I_{Kr} and cause drug-induced LQTS.

Most drugs that block I_{Kr} bind to the inner aqueous vestibule at the Y652 and F656 residue in S6 of the Kv11.1a α-subunit [54,55]. Ficker and colleagues (2002) found that engineering amino acid substitutions at F656 reduced both the affinity of the drug blocker of I_{Kr} and the pharmacological correction of Class 2 LQT2 channel proteins [45]. Conversely, engineering certain amino acid substitutions at the Y652 residue (e.g., Y652C) mitigated the trafficking-deficient phenotypes for several different Class 2 LQT2 mutations [56]. These latter data are an example of intragenic suppression, whereby variants at Y652 compensate for the primary misfolding defect caused by disease-causing Class 2 LQT2 mutation. Taken together, the results demonstrate that cleanly separating drug block and pharmacological correction might not be possible.

I_{Kr} activators increase the open probability of Kv11.1a channels at the cell surface membrane and might be used as pharmacotherapy for LQT2 [57]. However, they could lead to abnormal shortening of the QT interval and increase the risk for deadly arrhythmias like ventricular fibrillation. Qile and colleagues (2020) determined whether the functional expression of Class 2 LQT2 Kv11.1 channels could be increased by culturing cells in I_{Kr} blockers and the I_{Kr} activator LUF7244 [58,59]. LUF7244 is an allosteric I_{Kr} activator that counteracts I_{Kr} blocker induced arrhythmias in dogs. Heterologous expression studies showed that this strategy of combining drugs that increase trafficking and open probability of Kv11.1a channels was effective at increasing the functional expression of Class 2 LQT2 channel proteins at the cell surface membrane (even without drug wash out). This raises the possibility that I_{Kr} blockers and I_{Kr} activators could be used together to increase mutant channel trafficking and function at the cell surface membrane.

Many different congenital diseases are caused by mutations that disrupt protein folding and trafficking [60]. One of the best studied diseases caused by mutations that disrupt ion channel trafficking is cystic fibrosis [61–63]. Cystic fibrosis is caused by LOF mutations in the *cystic fibrosis transmembrane conductance regulator (CFTR)*. *CFTR* encodes an ABC transporter-class Cl$^-$ channel and most cases of cystic fibrosis cases are linked to the deletion mutation (F508del-CFTR), which increases CFTR channel protein misfolding and decreases its trafficking out of the ER [64,65]. Recent investigational therapeutic strategies to treat CFTR include the drugs that can facilitate the native folding and membrane expression of CFTR channels (lumacaftor), as well as drugs that increase the open probability of CFTR channels (ivacaftor) [66,67].

Mehta and colleagues (2018) determined whether lumacaftor also improves the trafficking of Class 2 LQT2 channel proteins [68]. These studies began with in vitro testing in inducible pluripotent stem cell-derived cardiomyocytes (iPSC-CMs) generated from patients that have Class 2 LQT2 mutations. Incubating iPSC-CMs in lumacaftor shortened the field potential duration, which is the iPSC-CM rough equivalent to the QT interval [68]. Schwartz and colleagues [69] tested whether lumacaftor and ivacaftor could normalize the QT interval in several LQT2 patients who harbor the same Class 2 LQT2 mutation as in iPSC-CMs studies. The patients treated with these drugs showed shortening in their QTc interval, but the absolute magnitude was less than expected based on the iPSC-CM data and the patients suffered some undesirable side effects. This discrepancy could be caused by differences to drug delivery to iPSC-CMs in a dish vs. an intact organ, the phenotypic immaturity of iPSC-CMs, and/or the observation that iPSC-CMs express a relatively large I_{Kr} [70]. Additional considerations could be the systemic changes that these drugs might cause on physiological variables important for maintaining the QT interval. The authors concluded that the use of this drug combination for treating LQT2 patients is premature until larger patient populations can be studied.

Although improving the trafficking of Class 2 Kv11.1a channel proteins represents an attractive therapeutic strategy, it might be impractical because of the large number of different mutations and their diverse trafficking and biophysical molecular phenotypes (for review see [71]). It is clear that pharmacological correction strategies improve the trafficking of certain mutations but not others. Even if pharmacological correction of Class 2 Kv11.1a channel proteins can be done in patients, there is no guarantee that the corrected mutant channel proteins will function normally to mitigate the clinical phenotype. Indeed, increasing the trafficking of mutant channel proteins that also severely disrupt Kv11.1a channel gating and/or permeation could worsen the disease.

1.4. KCNH2 Variants of Uncertain Significance (VUS)

Since we already have several effective therapies to currently treat patients with LQT2, then genetic screening of large patient populations could facilitate the early identification and prophylactic treatment [72,73]. A challenge is that most genetic tests and Whole Genome/Exome Sequencing identify novel, rare *KCNH2* missense variants of uncertain physiological significance (VUS). The vast majority of these VUS are likely neutral because the allelic frequencies for a VUS in *KCNH2* far outpaces the true incidence of the disease [12,74–77]. Therefore, positive genetic tests for VUS are not actionable.

One might think that laboratories experienced in genetic testing, ion channels, and cardiac arrhythmias would be good at predicting which LQT2-linked VUS are potentially pathogenic. However, this is proving to be extremely difficult. Van Driest et al. (2016) [78] found that, not only was there a discordance between different laboratories in designating a LQTS-linked VUS as pathogenic, but also that certain VUS identified as pathogenic did not associate with autosomal dominant LQTS. An incomplete understanding in the probabilistic nature of genetic testing, coupled with unexpected high allelic frequencies of neutral VUS, has already contributed to LQTS misdiagnoses and unnecessary treatment in patients, including invasive procedures with potential complications [52,79]. The current clinical recommendations now limit genetic screening to phenotypically positive LQT2 patients to prevent the misdiagnosis and over treatment of patients [12,80]. Since one of the presenting symptoms of LQT2 is sudden cardiac death, patients would benefit from the development of reliable gene-first strategies that identify LOF *KCNH2* VUS before they experience a life-threatening arrhythmia.

A joint consensus guideline document from the American College of Medical Genetics and Genomics and the Association of Molecular Pathology provides some guidance for the interpretation of sequence variants [81]. The rubric can be used to classify variants identified in Mendelian-linked disease alleles as pathogenic, likely pathogenic, uncertain significance, likely benign, and benign (Figure 4A). The rubric to classify variants uses population data (mean allelic frequency), computational data, functional data, and segregation data. It can be useful, but without functional data, it does not inform how predicted pathogenic or likely pathogenic mutations affect function. Additionally, it does not mention the possibility of identifying genetic variants that act as genetic modifiers. These are *KCNH2* variants that do not cause autosomal dominant LQT2 but rather modify the risk of having LQTS in response to disease, drugs, or the presence of genetic variants in other LQTS-susceptibility genes. We expect that advances in computational and structural biology will help not only determine how specific variants impact the structure but also the function for a protein of interest (Figure 4B,C). Ideally, computer simulations of known variants in Kv11.1a protein structures can be used to identify structural features toward predicting which newly identified VUS are neutral, genetic modifiers, or result in a LOF (or gain of function) by a Class 2, 3, or 4 mechanism.

In recent years, bioinformatics and protein-structure prediction approaches have leveraged primary structural information to infer structural or functional consequences of protein variants [82]. Disappointingly, while the PAS, pore and CNBD appear to harbor the majority of LQT2-linked variants, few correlations between the clinical severity and nature of their loss-of-function defects have emerged. Dozens of missense variants identified within or near these sequences present no obvious functional or clinical phenotype [83,84]. This severely undermines the utility of using sequence position alone as a way to infer LOF Kv11.1a variants.

Extending beyond analyses of the KCNH2 primary sequence alone, Anderson and colleagues provided considerable insights into how mutation-specific differences in the physicochemical properties and location of missense LQT2 variants in the PAS domain impact trafficking [85]. In this study, an in vitro assay reporting on protein solubility demonstrated that Class 2 LQT2 mutations in the PAS domain were considerably less soluble than their WT and trafficking-permissive counterparts. Further, the severity of trafficking defects loosely correlated with their relative insolubility as measured by the assay when compared to the WT Kv11.1a PAS domain. Interestingly, variants that undergo pharmacologic correction with I_{Kr} blockers were generally only mildly insoluble. It is tempting to speculate that mutant Kv11.1a channel proteins that undergo pharmacological correction generate relatively minor structural perturbations.

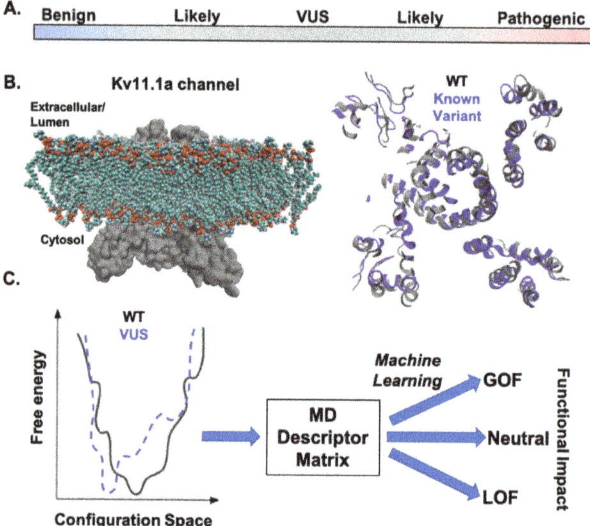

Figure 4. (**A**). The American College of Medical Genetics and Genomics and the Association of Molecular Pathology provides a rubric that can be used to classify variants identified in Mendelian-linked disease alleles as pathogenic, likely pathogenic, uncertain significance, likely benign, and benign. (**B**). Shown is an example of how molecular dynamics (MD) simulations could be used to determine the structural impact of known and newly identified variants on Kv11.1a channel configurations. (**C**). MD simulations could be used to relate structural configurations to thermodynamic changes, like those controlling folding, that cause a LOF, gain of function (GOF) or neutral phenotype. Such predictions could be compiled for known variants into a descriptor matrix to train machine learning approaches for de novo prediction of the structural and functional impact of newly identified VUS in Kv11.1a channels. Over time, the sensitivity of these approaches might be fine-tuned to determine which VUS act as genetic modifiers.

Ascribing physicochemical properties such as hydrophobicity based on single amino acid substitutions and primary sequence alone to predict the impact of an individual mutation on Kv11.1a function is challenging. Bioinformatics approaches have nonetheless identified statistically significant associations between folding free energies and genetic sequences [86,87]. Differences in Kv11.1a channel folding free energies can serve as suggestive indicators of misfolding and defective trafficking, as shown in Anderson et al. [25] It is important to remember that the effects of missense variants are not restricted to only the regions of the protein that are adjacent in primary sequence, but they can act allosterically, as they couple to and depend on the structural configuration of the entire protein. The interdependence of adjacent and nonadjacent sequence information in determining the impact of a mutation on channel function underscores the need for incorporating structural information for predicting the functional impact of individual variants [88].

Effects of mutations on Kv11.1a protein function can be interpreted on the basis of how they impact secondary (alpha helix, random coil, beta sheets), tertiary (protein and subdomain folding) and quaternary (protein/protein associations) factors. Myriad algorithms have been introduced to predict protein folding tendencies for forming canonical secondary structures, identify protein folds based on sequence similarities to other protein structures deposited in the Protein Databank, and predict potential protein–protein interaction interfaces [82,89–91]. Nonetheless, a gold standard for surmising structure/function relationships in proteins and how they might be perturbed by missense mutations is interpreting atomistic-scale structural data of intact proteins, when available. To date, such information for Kv11.1a is incomplete and comprises only Angstrom resolution models of its PAS, EAG and CNBD domains but not of the whole channel [92–94]. However, a significant advance in this regard is the recent availability of sub-nanometer resolution structural data of the Kv11.1a channel via cryo-electron microscopy [27]. This structure has been critical in the initial stages of interpreting how known LQT2 mutations, as well as newly identified KCNH2 VUS, impact Kv11.1a properties and its function.

Computational simulation techniques including homology-modeling and molecular dynamics simulations (MDS) are among the most popular tools for predicting the structural bases of protein function and potential changes in function from mutations. Homology models are used to predict the likely three-dimensional structure of a primary sequence using structural data available in homologous proteins as templates [95]. The accuracy of these approaches are highly dependent on the quality of the template data and sequence identity. Molecular simulations generally use physics- or knowledge-based scoring functions to estimate interaction potentials between amino acids or atoms, which are used to predict three-dimensional protein structures or how they move under those potentials [96]. Recently, cryo-electron microscopy based models and MDS simulations of Kv11.1a channel proteins have been used to elucidate a gain-of-function (GOF) mutation that impacts the channel's selectivity filter [97]. This study was a first-of-its-kind approach for linking KCNH2 genomic information to a functional outcome using structurally-realistic models of the channel and computer simulations.

While the use of molecular simulations of the Kv11.1a channel represent a significant and innovative step toward predicting the pathogenicity of KCNH2 variants, several advancements are still needed to fully realize this potential. Cryo-electron data are relatively low resolution (sub nanometer) compared to crystallography- or NMR-derived structures (Angstrom) limiting the ability to conclusively link a mutation with its mechanism of dysfunction. While a recent advancement in cryo-electron microscopy has resolved individual atoms in apoferritin [98], it is unclear when such resolution gains may be accomplished for membrane channels like Kv11.1a. In the meantime, continued improvements in the solubilization and crystallization of membrane proteins including ion channels will be required to approach atomistic resolution images via X-ray crystallography, as it has been done in recent years with voltage-gated Na^+ channels [99]. Additionally, efficiently implementation of MDS will benefit from force fields that include the effects of electrostatic polarization as the local electric field changes [100].

MDS based on these structural data, regardless of the structural resolution, remain intractable for modeling molecular events at microsecond or longer timescales relevant for biological function. Recent advances in specialized hardware, including graphics processing units and chip architectures, have begun to approach these timescales and have been used to probe mechanisms of channel folding and function (reviewed in [101]) [102,103]. In this regard, molecular simulations leveraging these technological advancements could enable predictions of potential trafficking behavior from first principles for the vast majority of known mutations and future KCNH2 VUS. To realize this outcome, computer simulations of these processes will require defining and determining the structural regions of the Kv11.1a channel implicated in defective trafficking, such as its PAS domain, as well as highlighting properties like solubility that correlate with LOF. Similarly, the resolution of complexes between the channel and chaperones including heat shock proteins may further yield important mechanistic clues for how missense variants impact ER quality control and trafficking [31,32].

Naturally, a compelling motivation for using computational techniques to determine relationships between genetic information and Kv11.1a dysfunction is to provide predictive information that is superior to genetic screening but not as labor intensive, expensive, or slow as functional studies using in vitro or iPSC-CMs. Achieving this prediction capacity remains the critical barrier to using genetic screening more effectively in a clinical setting. If this goal of predictive power is met, there will still be a secondary barrier in ensuring that the predictive information is easy to obtain and interpret in clinical practice. Most clinicians would obviously not have the ability to perform advance bioinformatics analyses and computer simulations in their offices! Here statistical tools and machine learning approaches may provide the means to distill insights from detailed bioinformatics analyses and simulations into searchable databases. Such efforts have already been documented for machine learning and data mining applied to diabetes research, as well as other genome-wide association studies [104,105].

2. Summary

It has been almost 25 years since KCNH2 was linked to LQTS. Since then we have learned that most LQT2-linked mutations likely decrease IKr by decreasing Kv11.1 channel synthesis or trafficking. The identification of drugs that can improve mutant channel trafficking and increase the open probability could represent a novel therapeutic approach. Perhaps an equally or even more important challenge is harnessing genetic analysis to help identify new LQT2 patients before they suffer a life-threatening arrhythmic event. We expect that recent advances in solving Kv11.1a channel structures, computational simulations, and bio-informatics will improve our ability to distinguish neutral VUS from those that confer a high-risk for LQT2.

Author Contributions: M.O., D.E.B., E.A.S., C.S.E., C.L.A., C.T.J., B.S., K.I., P.M.K.-H. and B.P.D. played a role in the development, drafting and/or revision of the review article. All authors have read and agreed to the published version of the manuscript.

Funding: This work was supported by the Maximizing Investigators' Research Award (MIRA) (R35) from the National Institute of General Medical Sciences (NIGMS) of the National Institutes of Health (NIH) under grant number R35GM124977.

Conflicts of Interest: The authors declare no conflict of interest.

References

1. Moss, A.J. Long QT syndrome. *JAMA* **2003**, *289*, 2041–2044. [CrossRef] [PubMed]
2. Schwartz, P.J.; Moss, A.J.; Vincent, G.M.; Crampton, R.S. Diagnostic criteria for the long Qt syndrome. An update. *Circulation* **1993**, *88*, 782–784. [CrossRef]
3. El-Sherif, N.; Turitto, G.; Boutjdir, M. Congenital Long QT syndrome and torsade de pointes. *Ann. Noninvasive Electrocardiol.* **2017**, *22*, 1–11. [CrossRef] [PubMed]
4. Schwartz, P.J.; Stramba-Badiale, M.; Crotti, L.; Pedrazzini, M.; Besana, A.; Bosi, G.; Gabbarini, F.; Goulene, K.; Insolia, R.; Mannarino, S.; et al. Prevalence of the congenital long-QT syndrome. *Circulation* **2009**, *120*, 1761–1767. [CrossRef]
5. Sanguinetti, M.C.; Jiang, C.; Curran, M.E.; Keating, M.T. A mechanistic link between an inherited and an acquired cardiac arrhythmia: HERG encodes the IKr potassium channel. *Cell* **1995**, *81*, 299–307. [CrossRef]
6. Trudeau, M.; Warmke, J.; Ganetzky, B.; Robertson, G. HERG, a human inward rectifier in the voltage-gated potassium channel family. *Science* **1995**, *269*, 92–95. [CrossRef]
7. Wang, Q.; Curran, M.; Splawski, I.; Burn, T.C.; Millholland, J.; VanRaay, T.; Shen, J.; Timothy, K.; Vincent, G.; De Jager, T.; et al. Positional cloning of a novel potassium channel gene: KVLQT1 mutations cause cardiac arrhythmias. *Nat. Genet.* **1996**, *12*, 17–23. [CrossRef]
8. Donger, C.; Denjoy, I.; Berthet, M.; Neyroud, N.; Cruaud, C.; Bennaceur, M.; Chivoret, G.; Schwartz, K.; Coumel, P.; Guicheney, P. KVLQT1 C-terminal missense mutation causes a forme fruste long-QT syndrome. *Circulation* **1997**, *96*, 2778–2781. [CrossRef]
9. Curran, M.E.; Splawski, I.; Timothy, K.W.; Vincen, G.; Green, E.D.; Keating, M.T. A molecular basis for cardiac arrhythmia: HERG mutations cause long QT syndrome. *Cell* **1995**, *80*, 795–803. [CrossRef]
10. Wang, Q.; Shen, J.; Splawski, I.; Atkinson, D.; Li, Z.; Robinson, J.L.; Moss, A.J.; Towbin, J.A.; Keating, M.T. SCN5A mutations associated with an inherited cardiac arrhythmia, long QT syndrome. *Cell* **1995**, *80*, 805–811. [CrossRef]
11. Adler, A.; Novelli, V.; Amin, A.S.; Abiusi, E.; Care, M.; Nannenberg, E.A.; Feilotter, H.; Amenta, S.; Mazza, D.; Bikker, H.; et al. An international, multicentered, evidence-based reappraisal of genes reported to cause congenital long QT Syndrome. *Circulation* **2020**, *141*, 418–428. [CrossRef] [PubMed]

12. Priori, S.G.; Wilde, A.A.; Horie, M.; Cho, Y.; Behr, E.R.; Berul, C.; Blom, N.; Brugada, J.; Chiang, C.E.; Huikuri, H.; et al. Executive Summary: Hrs/Ehra/Aphrs expert consensus statement on the diagnosis and management of patients with inherited primary arrhythmia syndromes. *Heart Rhythm* **2013**, *10*, 85–108. [CrossRef] [PubMed]

13. Schwartz, P.J.; Crotti, L.; Insolia, R. Long-Qt syndrome: From genetics to management. *Circ. Arrhythmia Electrophysiol.* **2012**, *5*, 868–877. [CrossRef] [PubMed]

14. Wang, Q.; Chen, Q.; Towbin, J.A. Genetics, molecular mechanisms and management of long QT syndrome. *Ann. Med.* **1998**, *30*, 58–65. [CrossRef]

15. Delisle, B.P.; Anson, B.D.; Rajamani, S.; January, C.T. Biology of cardiac Arrhythmias. *Circ. Res.* **2004**, *94*, 1418–1428. [CrossRef]

16. Smith, J.L.; Anderson, C.L.; Burgess, D.E.; Elayi, C.S.; January, C.T.; Delisle, B.P. Molecular pathogenesis of long QT syndrome type. *J. Arrhythmia* **2016**, *32*, 373–380. [CrossRef]

17. London, B.; Trudeau, M.C.; Newton, K.P.; Beyer, A.K.; Copeland, N.G.; Gilbert, D.J.; Jenkins, N.A.; Satler, C.A.; Robertson, G.A. Two isoforms of the mouse ether-a-go-go-related gene coassemble to form channels with properties similar to the rapidly activating component of the cardiac delayed rectifier K+ current. *Circ. Res.* **1997**, *81*, 870–878. [CrossRef]

18. Abbott, G.W.; Sesti, F.; Splawski, I.; Buck, M.E.; Lehmann, M.H.; Timothy, K.W.; Keating, M.T.; Goldstein, S.A. MiRP1 forms I Kr potassium channels with HERG and is associated with cardiac Arrhythmia. *Cell* **1999**, *97*, 175–187. [CrossRef]

19. Jonsson, M.; Van Der Heyden, M.A.; Van Veen, T.A. Deciphering hERG channels: Molecular basis of the rapid component of the delayed rectifier potassium current. *J. Mol. Cell. Cardiol.* **2012**, *53*, 369–374. [CrossRef]

20. Gong, Q.; Stump, M.R.; Dunn, A.R.; Deng, V.; Zhou, Z. Alternative splicing and Polyadenylation contribute to the generation of hERG1 C-terminal Isoforms. *J. Biol. Chem.* **2010**, *285*, 32233–32241. [CrossRef]

21. Kupershmidt, S.; Snyders, D.J.; Raes, A.; Roden, D.M. A K+ channel splice variant common in human heart lacks a C-terminal domain required for expression of rapidly activating delayed rectifier current. *J. Biol. Chem.* **1998**, *273*, 27231–27235. [CrossRef] [PubMed]

22. Abbott, G.W. KCNE2 and the K+ channel. *Channels* **2012**, *6*, 1–10. [CrossRef] [PubMed]

23. Roberts, J.D.; Krahn, A.D.; Ackerman, M.J.; Rohatgi, R.K.; Moss, A.J.; Nazer, B.; Tadros, R.; Gerull, B.; Sanatani, S.; Wijeyeratne, Y.D.; et al. Loss-of-Function KCNE2 Variants. *Circ. Arrhythmia Electrophysiol.* **2017**, *10*, 1–11. [CrossRef] [PubMed]

24. Anderson, C.L.; Delisle, B.P.; Anson, B.D.; Kilby, J.A.; Will, M.L.; Tester, D.J.; Gong, Q.; Zhou, Z.; Ackerman, M.J.; January, C.T. Most Lqt2 mutations reduce Kv11.1 (Herg) current by a class 2 (Trafficking-Deficient) mechanism. *Circulation* **2006**, *113*, 365–373. [CrossRef] [PubMed]

25. Anderson, C.L.; Kuzmicki, C.E.; Childs, R.R.; Hintz, C.J.; Delisle, B.P.; January, C.T. Large-scale mutational analysis of Kv11.1 reveals molecular insights into type 2 long QT syndrome. *Nat. Commun.* **2014**, *5*, 5535–5548. [CrossRef]

26. Hille, B. *Ion Channels of Excitable Membranes*, 3rd ed.; Sinauer Associates, Inc.: Sunderland, MA, USA, 2001.

27. Wang, W.; MacKinnon, R. Cryo-EM structure of the open human ether-à-go-go related K+ channel hERG. *Cell* **2017**, *169*, 422–430. [CrossRef]

28. Schulteis, C.T.; Nagaya, N.; Papazian, D.M. Subunit folding and assembly steps are interspersed during shaker potassium channel biogenesis. *J. Biol. Chem.* **1998**, *273*, 26210–26217. [CrossRef]

29. Ellgaard, L.; Helenius, A. Quality control in the endoplasmic reticulum. *Nat. Rev. Mol. Cell Biol.* **2003**, *4*, 181–191. [CrossRef]

30. Zhou, Z.; Gong, Q.; Ye, B.; Fan, Z.; Makielski, J.C.; Robertson, G.A.; January, C.T. Properties of HERG channels stably expressed in HEK 293 cells studied at physiological temperature. *Biophys. J.* **1998**, *74*, 230–241. [CrossRef]

31. Foo, B.; Williamson, B.; Young, J.C.; Lukacs, G.L.; Shrier, A. hERG quality control and the long QT syndrome. *J. Physiol.* **2016**, *594*, 2469–2481. [CrossRef]

32. Ficker, E.; Dennis, A.T.; Wang, L.; Brown, A.M. Role of the cytosolic chaperones Hsp70 and Hsp90 in maturation of the cardiac potassium channel hERG. *Circ. Res.* **2003**, *92*, 87–100. [CrossRef] [PubMed]

33. Delisle, B.P.; Underkofler, H.A.S.; Moungey, B.M.; Slind, J.K.; Kilby, J.A.; Best, J.M.; Foell, J.D.; Balijepalli, R.C.; Kamp, T.J.; January, C.T. Small GTPase determinants for the Golgi processing and plasmalemmal expression of human ether-a-go-go Related (hERG) K+Channels. *J. Biol. Chem.* **2008**, *284*, 2844–2853. [CrossRef] [PubMed]

34. Gong, Q.; Anderson, C.L.; January, C.T.; Zhou, Z. Role of glycosylation in cell surface expression and stability of HERG potassium channels. *Am. J. Physiol. Circ. Physiol.* **2002**, *283*, 77–84. [CrossRef] [PubMed]

35. Foo, B.; Barbier, C.; Guo, K.; Vasantharuban, J.; Lukacs, G.L.; Shrier, A. Mutation-specific peripheral and ER quality control of hERG channel cell-surface expression. *Sci. Rep.* **2019**, *9*, 10113–10121. [CrossRef]

36. Kanner, S.A.; Jain, A.; Colecraft, H.M. Development of a high-throughput flow cytometry assay to monitor defective trafficking and rescue of Long QT2 Mutant hERG channels. *Front. Physiol.* **2018**, *9*, 397–407. [CrossRef]

37. Perry, M.D.; Ng, C.A.; Mann, S.A.; Sadrieh, A.; Imtiaz, M.; Hill, A.P.; Vandenberg, J.I. Getting to the heart of Herg K(+) channel Gating. *J. Physiol.* **2015**, *593*, 2575–2585. [CrossRef]

38. Gong, Q.; Zhang, L.; Vincent, G.M.; Horne, B.D.; Zhou, Z. Nonsense mutations in hERG cause a decrease in mutant mRNA transcripts by nonsense-mediated mRNA decay in human long-QT syndrome. *Circulation* **2007**, *116*, 17–24. [CrossRef]

39. Stump, M.R.; Gong, Q.; Zhou, Z. LQT2 nonsense mutations generate trafficking defective NH2-terminally truncated channels by the reinitiation of translation. *Am. J. Physiol. Circ. Physiol.* **2013**, *305*, H1397–H1404. [CrossRef]

40. Splawski, I.; Shen, J.; Timothy, K.W.; Lehmann, M.H.; Priori, S.; Robinson, J.L.; Moss, A.J.; Schwartz, P.J.; Towbin, J.A.; Vincent, G.M.; et al. Spectrum of mutations in long-QT syndrome genes. KVLQT1, HERG, SCN5A, KCNE1, and KCNE. *Circulation* **2000**, *102*, 1178–1185. [CrossRef]

41. Zhou, Z.; Gong, Q.; Epstein, M.L.; January, C.T. HERG channel dysfunction in human Long QT syndrome. *J. Biol. Chem.* **1998**, *273*, 21061–21066. [CrossRef]

42. Chen, J.; Zou, A.; Splawski, I.; Keating, M.T.; Sanguinetti, M.C. Long QT syndrome-associated mutations in the Per-Arnt-Sim (PAS) domain of HERG potassium channels accelerate channel deactivation. *J. Biol. Chem.* **1999**, *274*, 10113–10118. [CrossRef] [PubMed]

43. Wakana, Y.; Takai, S.; Nakajima, K.-I.; Tani, K.; Yamamoto, A.; Watson, P.; Stephens, D.; Hauri, H.-P.; Tagaya, M. Bap31 is an itinerant protein that moves between the Peripheral Endoplasmic Reticulum (ER) and a Juxtanuclear compartment Related to ER-associated degradation. *Mol. Biol. Cell* **2008**, *19*, 1825–1836. [CrossRef] [PubMed]

44. Smith, J.L.; Reloj, A.R.; Nataraj, P.S.; Bartos, D.C.; Schroder, E.A.; Moss, A.J.; Ohno, S.; Horie, M.; Anderson, C.L.; January, C.T.; et al. Pharmacological correction of long QT-linked mutations in KCNH2 (hERG) increases the trafficking of Kv11.1 channels stored in the transitional endoplasmic reticulum. *Am. J. Physiol. Physiol.* **2013**, *305*, 919–930. [CrossRef] [PubMed]

45. Ficker, E.; Obejero-Paz, C.A.; Zhao, S.; Brown, A.M. The binding site for channel blockers that rescue misprocessed human long QT syndrome type 2ether-a-gogo-related gene (HERG) Mutations. *J. Biol. Chem.* **2001**, *277*, 4989–4998. [CrossRef] [PubMed]

46. Furutani, M.; Trudeau, M.C.; Hagiwara, N.; Seki, A.; Gong, Q.; Zhou, Z.; Imamura, S.; Nagashima, H.; Kasanuki, H.; Takao, A.; et al. Novel mechanism associated with an inherited cardiac arrhythmia: Defective protein trafficking by the mutant HERG (G601S) potassium channel. *Circulation* **1999**, *99*, 2290–2294. [CrossRef]

47. Hall, A.R.; Anderson, C.L.; Smith, J.L.; Mirshahi, T.; Elayi, C.S.; January, C.T.; Delisle, B.P. Visualizing mutation-specific differences in the trafficking-deficient Phenotype of Kv11.1 proteins linked to long QT syndrome type. *Front. Physiol.* **2018**, *9*, 584–595. [CrossRef] [PubMed]

48. Ficker, E.; Thomas, D.; Viswanathan, P.C.; Dennis, A.T.; Priori, S.G.; Napolitano, C.; Memmi, M.; Wible, B.A.; Kaufman, E.S.; Iyengar, S.K.; et al. Novel characteristics of a misprocessed mutant HERG channel linked to hereditary long QT syndrome. *Am. J. Physiol. Circ. Physiol.* **2000**, *279*, 1748–1756. [CrossRef]

49. Hirsch, C.; Ploegh, H.L. Intracellular targeting of the proteasome. *Trends Cell Biol.* **2000**, *10*, 268–272. [CrossRef]

50. Gong, Q.; Keeney, D.R.; Molinari, M.; Zhou, Z. Degradation of trafficking-defective long QT syndrome type II mutant channels by the ubiquitin-proteasome Pathway. *J. Biol. Chem.* **2005**, *280*, 19419–19425. [CrossRef]

51. Ficker, E.; Dennis, A.T.; Obejero-Paz, C.A.; Castaldo, P.; Taglialatela, M.; Brown, A.M. Retention in the endoplasmic reticulum as a mechanism of dominant-negative current suppression in human long QT syndrome. *J. Mol. Cell. Cardiol.* **2000**, *32*, 2327–2337. [CrossRef]

52. Schwartz, P.J.; Spazzolini, C.; Silvia, P.G.; Crotti, L.; Vicentini, A.; Landolina, M.; Gasparini, M.; Wilde, A.A.M.; Knops, R.E.; Denjoy, I.; et al. Who are the long-QT syndrome patients who receive an implantable cardioverter-defibrillator and what happens to them? *Circulation* **2010**, *122*, 1272–1282. [CrossRef] [PubMed]

53. Zhou, Z.; Gong, Q.; January, C.T. Correction of defective protein trafficking of a mutant HERG potassium channel in human long QT syndrome. Pharmacological and temperature effects. *J. Biol. Chem.* **1999**, *274*, 31123–31126. [CrossRef] [PubMed]

54. Lees-Miller, J.P.; Duan, Y.; Teng, G.Q.; Duff, H.J. Molecular determinant of high-affinity dofetilide binding to HERG1 expressed in Xenopus oocytes: Involvement of S6 sites. *Mol. Pharmacol.* **2000**, *57*, 367–374. [PubMed]

55. Mitcheson, J.S.; Chen, J.; Lin, M.; Culberson, C.; Sanguinetti, M.C. A structural basis for drug-induced long QT syndrome. *Proc. Natl. Acad. Sci. USA* **2000**, *97*, 12329–12333. [CrossRef] [PubMed]

56. Delisle, B.P.; Slind, J.K.; Kilby, J.A.; Anderson, C.L.; Anson, B.D.; Balijepalli, R.C.; Tester, D.J.; Ackerman, M.J.; Kamp, T.J.; January, C.T. Intragenic suppression of trafficking-defective KCNH2 channels associated with long QT syndrome. *Mol. Pharmacol.* **2005**, *68*, 233–240. [CrossRef] [PubMed]

57. Sanguinetti, M.C. HERG1 channel agonists and cardiac arrhythmia. *Curr. Opin. Pharmacol.* **2013**, *15*, 22–27. [CrossRef]

58. Qile, M.; Beekman, H.D.; Sprenkeler, D.J.; Houtman, M.J.; Van Ham, W.B.; Stary-Weinzinger, A.; Beyl, S.; Hering, S.; Berg, D.V.D.; De Lange, E.C.M.; et al. LUF7244, an allosteric modulator/activator of K v 11.1 channels, counteracts dofetilide-induced torsades de pointes arrhythmia in the chronic atrioventricular block dog model. *Br. J. Pharmacol.* **2019**, *176*, 3871–3885. [CrossRef]

59. Qile, M.; Ji, Y.; Golden, T.D.; Houtman, M.J.; Romunde, F.; Fransen, D.; Van Ham, W.B.; Ijzerman, A.P.; January, C.T.; Heitman, L.H.; et al. LUF7244 plus Dofetilide rescues aberrant Kv11.1 trafficking and produces functional IKv11. *Mol. Pharmacol.* **2020**, *97*, 355–364. [CrossRef]

60. Aridor, M.; Hannan, L.A. Traffic Jam: A compendium of human diseases that affect intracellular transport processes. *Traffic* **2000**, *1*, 836–851. [CrossRef]

61. Denning, G.M.; Anderson, M.P.; Amara, J.F.; Marshall, J.; Smith, A.E.; Welsh, M.J. Processing of mutant cystic fibrosis transmembrane conductance regulator is temperature-sensitive. *Nature* **1992**, *358*, 761–764. [CrossRef]

62. Welsh, M.J.; Smith, A.E. Molecular mechanisms of CFTR chloride channel dysfunction in cystic fibrosis. *Cell* **1993**, *73*, 1251–1254. [CrossRef]

63. McClure, M.L.; Barnes, S.; Brodsky, J.L.; Sorscher, E.J. Trafficking and function of the cystic fibrosis transmembrane conductance regulator: A complex network of posttranslational modifications. *Am. J. Physiol. Cell. Mol. Physiol.* **2016**, *311*, 719–733. [CrossRef] [PubMed]

64. Serohijos, A.W.R.; Hegedűs, T.; Aleksandrov, A.A.; He, L.; Cui, L.; Dokholyan, N.V.; Riordan, J.R. Phenylalanine-508 mediates a cytoplasmic-membrane domain contact in the CFTR 3D structure crucial to assembly and channel function. *Proc. Natl. Acad. Sci. USA* **2008**, *105*, 3256–3261. [CrossRef] [PubMed]

65. Lukacs, G.L.; Chang, X.B.; Bear, C.; Kartner, N.; Mohamed, A.; Riordan, J.R.; Grinstein, S. The delta F508 mutation decreases the stability of cystic fibrosis transmembrane conductance regulator in the plasma membrane. Determination of functional half-lives on transfected cells. *J. Biol. Chem.* **1993**, *268*, 21592–21598.

66. Van Goor, F.; Hadida, S.; Grootenhuis, P.D.J.; Burton, B.; Stack, J.H.; Straley, K.S.; Decker, C.J.; Miller, M.; McCartney, J.; Olson, E.R.; et al. Correction of the F508del-CFTR protein processing defect in vitro by the investigational drug VX-809. *Proc. Natl. Acad. Sci. USA* **2011**, *108*, 18843–18848. [CrossRef]

67. Connett, G. Lumacaftor-ivacaftor in the treatment of cystic fibrosis: Design, development and place in therapy. *Drug Des. Dev. Ther.* **2019**, *13*, 2405–2412. [CrossRef]

68. Mehta, A.; Ramachandra, C.J.A.; Singh, P.; Chitre, A.; Lua, C.H.; Mura, M.; Crotti, L.; Wong, P.; Schwartz, P.J.; Gnecchi, M.; et al. Identification of a targeted and testable antiarrhythmic therapy for long-QT syndrome type 2 using a patient-specific cellular model. *Eur. Hear. J.* **2017**, *39*, 1446–1455. [CrossRef]

69. Schwartz, P.J.; Gnecchi, M.; Dagradi, F.; Castelletti, S.; Parati, G.; Spazzolini, C.; Sala, L.; Crotti, L. From patient-specific induced pluripotent stem cells to clinical translation in long QT syndrome type. *Eur. Hear. J.* **2019**, *40*, 1832–1836. [CrossRef]

70. Braam, S.; Tertoolen, L.; Casini, S.; Matsa, E.; Lu, H.; Teisman, A.; Passier, R.; Denning, C.; Gallacher, D.; Towart, R.; et al. Repolarization reserve determines drug responses in human pluripotent stem cell derived cardiomyocytes. *Stem Cell Res.* **2013**, *10*, 48–56. [CrossRef]

71. Kaufman, E.S.; Ficker, E. Is restoration of intracellular trafficking clinically feasible in the long QT syndrome?: The example of HERG mutations. *J. Cardiovasc. Electrophysiol.* **2003**, *14*, 320–322. [CrossRef]

72. Shah, S.H.; Arnett, D.; Houser, S.R.; Ginsburg, G.S.; Macrae, C.; Mital, S.; Loscalzo, J.; Hall, J.L. Mhs opportunities for the cardiovascular community in the precision medicine initiative. *Circulation* **2016**, *133*, 226–231. [CrossRef] [PubMed]

73. National Research Council Committee on, A. Framework for Developing a New Taxonomy of Disease. The National Academies Collection: Reports Funded by National Institutes of Health. In *Toward Precision Medicine: Building a Knowledge Network for Biomedical Research and a New Taxonomy of Disease*; National Academies Press: Washington, DC, USA, 2011.

74. Kapa, S.; Tester, D.J.; Salisbury, B.A.; Harris-Kerr, C.; Pungliya, M.S.; Alders, M.; Wilde, A.A.M.; Ackerman, M.J. Genetic testing for long-QT syndrome: Distinguishing pathogenic mutations from benign variants. *Circulation* **2009**, *120*, 1752–1760. [CrossRef] [PubMed]

75. Taggart, N.W.; Haglund, C.M.; Tester, D.J.; Ackerman, M.J. Diagnostic miscues in congenital long-QT syndrome. *Circulation* **2007**, *115*, 2613–2620. [CrossRef] [PubMed]

76. Schwartz, P.J.; Ackerman, M.J.; George, A.L.; Wilde, A.A. Impact of genetics on the clinical management of Channelopathies. *J. Am. Coll. Cardiol.* **2013**, *62*, 169–180. [CrossRef] [PubMed]

77. Tester, D.J.; Schwartz, P.J.; Ackerman, M.J. Congenital long QT syndrome. *Electr. Dis. Heart* **2013**, *3*, 439–468. [CrossRef]

78. Van Driest, S.L.; Wells, Q.S.; Stallings, S.; Bush, W.S.; Gordon, A.; Nickerson, D.A.; Kim, J.H.; Crosslin, D.R.; Jarvik, G.P.; Carrell, D.S.; et al. Association of Arrhythmia-related genetic variants with phenotypes documented in electronic medical records. *JAMA* **2016**, *315*, 47–57. [CrossRef]

79. Gaba, P.; Bos, J.M.; Cannon, B.C.; Cha, Y.-M.; Friedman, P.A.; Asirvatham, S.J.; Ackerman, M.J. Implantable cardioverter-defibrillator explantation for overdiagnosed or overtreated congenital long QT syndrome. *Hear. Rhythm.* **2016**, *13*, 879–885. [CrossRef]

80. Ackerman, M.J.; Priori, S.G.; Willems, S.; Berul, C.; Brugada, R.; Calkins, H.; Camm, A.J.; Ellinor, P.T.; Gollob, M.; Hamilton, R.; et al. Hrs/Ehra expert consensus statement on the state of genetic testing for the Channelopathies and Cardiomyopathies. This document was developed as a partnership between the Heart Rhythm Society (HRS) and the European Heart Rhythm Association (EHRA). *Heart Rhythm* **2011**, *8*, 1308–1339. [CrossRef]

81. Richards, S.; Aziz, N.; Bale, S.; Bick, D.; Das, S.; Gastier-Foster, J.; Grody, W.W.; Hegde, M.; Lyon, E.; Spector, E.; et al. Standards and guidelines for the interpretation of sequence variants: A joint consensus recommendation of the American College of Medical Genetics and Genomics and the Association for Molecular Pathology. *Genet. Med.* **2015**, *17*, 405–423. [CrossRef]

82. Kuhlman, B.; Bradley, P. Advances in protein structure prediction and design. *Nat. Rev. Mol. Cell Biol.* **2019**, *20*, 681–697. [CrossRef]

83. Vanoye, C.G.; George, A.L. Decoding KCNH2 variants of unknown significance. *Hear. Rhythm.* **2020**, *17*, 501–502. [CrossRef] [PubMed]

84. Ng, C.-A.; Perry, M.D.; Liang, W.; Smith, N.J.; Foo, B.; Shrier, A.; Lukacs, G.L.; Hill, A.P.; Vandenberg, J. High-throughput phenotyping of heteromeric human ether-à-go-go-related gene potassium channel variants can discriminate pathogenic from rare benign variants. *Hear. Rhythm.* **2020**, *17*, 492–500. [CrossRef] [PubMed]

85. Anderson, C.L.; Routes, T.C.; Eckhardt, L.L.; Delisle, B.P.; January, C.T.; Kamp, T.J. A rapid solubility assay of protein domain misfolding for pathogenicity assessment of rare DNA sequence variants. *Genet. Med.* **2020**. [CrossRef] [PubMed]

86. Peng, Y.; Alexov, E. Investigating the linkage between disease-causing amino acid variants and their effect on protein stability and binding. *Proteins Struct. Funct. Bioinform.* **2016**, *84*, 232–239. [CrossRef] [PubMed]

87. Gyulkhandanyan, A.; Rezaie, A.R.; Roumenina, L.; Lagarde, N.; Fremeaux-Bacchi, V.; Miteva, M.A.; Villoutreix, B.O. Analysis of protein missense alterations by combining sequence- and structure-based methods. *Mol. Genet. Genom. Med.* **2020**, *8*, 1–28. [CrossRef] [PubMed]

88. Changeux, J.-P.; Christopoulos, A. Allosteric modulation as a unifying mechanism for receptor function and regulation. *Diabetes Obes. Metab.* **2017**, *19*, 4–21. [CrossRef]

89. Berman, H.M.; Westbrook, J.; Feng, Z.; Gilliland, G.; Bhat, T.N.; Weissig, H.; Shindyalov, I.N.; Bourne, P.E. The protein data bank. *Nucleic Acids Res.* **2000**, *28*, 235–242. [CrossRef]

90. Eisenhaber, F.; Persson, B.; Argos, P. Protein structure prediction: Recognition of primary, secondary, and tertiary structural features from amino acid sequence. *Crit. Rev. Biochem. Mol. Biol.* **1995**, *30*, 1–94. [CrossRef]

91. Sowmya, G.; Ranganathan, S. Protein-protein interactions and prediction: A comprehensive overview. *Protein Pept. Lett.* **2014**, *21*, 779–789. [CrossRef]

92. Ben-Bassat, A.; Giladi, M.; Haitin, Y. Structure of KCNH2 cyclic nucleotide-binding homology domain reveals a functionally vital salt-bridge. *J. Gen. Physiol.* **2020**, *152*, 1–12. [CrossRef]

Biomolecules **2020**, *10*, 1144

93. Muskett, F.W.; Thouta, S.; Thomson, S.J.; Bowen, A.; Stansfeld, P.J.; Mitcheson, J.S. Mechanistic insight into Humanether-à-go-go-related Gene (hERG) K+Channel deactivation Gating from the solution structure of the EAG domain. *J. Biol. Chem.* **2010**, *286*, 6184–6191. [CrossRef] [PubMed]
94. Adaixo, R.; Harley, C.A.; Castro-Rodrigues, A.F.; Morais-Cabral, J.H. Structural properties of PAS Domains from the KCNH Potassium Channels. *PLoS ONE* **2013**, *8*, 1–9. [CrossRef] [PubMed]
95. Lohning, A.E.; Levonis, S.M.; Williams-Noonan, B.; Schweiker, S.S. A Practical guide to molecular docking and homology modelling for medicinal chemists. *Curr. Top. Med. Chem.* **2017**, *17*, 2023–2040. [CrossRef]
96. Nurisso, A.; Daina, A.; Walker, R.C. A Practical introduction to molecular dynamics simulations: Applications to homology modeling. *Adv. Struct. Saf. Stud.* **2011**, *857*, 137–173. [CrossRef]
97. Miranda, W.E.; Demarco, K.R.; Guo, J.; Duff, H.J.; Vorobyov, I.; Clancy, C.E.; Noskov, S.Y. Selectivity filter modalities and rapid inactivation of the hERG1 channel. *Proc. Natl. Acad. Sci. USA* **2020**, *117*, 2795–2804. [CrossRef] [PubMed]
98. Man Yip, K.; Fischer, N.; Paknia, E.; Chari, A.; Stark, H. Breaking the next Cryo-Em resolution barrier—atomic resolution determination of proteins! *bioRxiv* **2020**, *2020*, 1–21. [CrossRef]
99. Sula, A.; Booker, J.; Ng, L.C.T.; Naylor, C.E.; DeCaen, P.G.; Wallace, B. The complete structure of an activated open sodium channel. *Nat. Commun.* **2017**, *8*, 14205–14224. [CrossRef]
100. Jing, Z.; Liu, C.; Cheng, S.Y.; Qi, R.; Walker, B.D.; Piquemal, J.-P.; Ren, P. Polarizable force fields for biomolecular simulations: Recent advances and applications. *Annu. Rev. Biophys.* **2019**, *48*, 371–394. [CrossRef]
101. Demarco, K.R.; Bekker, S.; Vorobyov, I. Challenges and advances in atomistic simulations of potassium and sodium ion channel gating and permeation. *J. Physiol.* **2018**, *597*, 679–698. [CrossRef]
102. Götz, A.W.; Williamson, M.J.; Xu, D.; Poole, D.; Le Grand, S.; Walker, R.C. Routine microsecond molecular dynamics simulations with AMBER on GPUs. Generalized born. *J. Chem. Theory Comput.* **2012**, *8*, 1542–1555. [CrossRef]
103. Shaw, D.E.; Grossman, J.P.; Bank, J.A.; Batson, B.; Butts, J.A.; Chao, J.C.; Deneroff, M.M.; Dror, R.O.; Even, A.; Fenton, C.H.; et al. Anton 2: Raising the Bar for Performance and Programmability in a Special-Purpose Molecular Dynamics Supercomputer. In Proceedings of the SC '14 International Conference for High Performance Computing, Networking, Storage and Analysis, New Orleans, LA, USA, 16–21 November 2014.
104. Kavakiotis, I.; Tsave, O.; Salifoglou, A.; Maglaveras, N.; Vlahavas, I.; Chouvarda, I. Machine learning and data mining methods in diabetes research. *Comput. Struct. Biotechnol. J.* **2017**, *15*, 104–116. [CrossRef] [PubMed]
105. Tam, V.; Patel, N.; Turcotte, M.; Bossé, Y.; Paré, G.; Meyre, D. Benefits and limitations of genome-wide association studies. *Nat. Rev. Genet.* **2019**, *20*, 467–484. [CrossRef] [PubMed]

 *biomolecules*

Review

Disease Associated Mutations in K_{IR} Proteins Linked to Aberrant Inward Rectifier Channel Trafficking

Eva-Maria Zangerl-Plessl [1,†], Muge Qile [2,†], Meye Bloothooft [2], Anna Stary-Weinzinger [1] and Marcel A. G. van der Heyden [2,*]

[1] Department of Pharmacology and Toxicology, University of Vienna, 1090 Vienna, Austria; eva-maria.zangerl@univie.ac.at (E.-M.Z.-P.); anna.stary@univie.ac.at (A.S.-W.)

[2] Department of Medical Physiology, Division of Heart & Lungs, University Medical Center Utrecht, 3584 CM Utrecht, The Netherlands; M.Qile@umcutrecht.nl (M.Q.); meye10@hotmail.com (M.B.)

* Correspondence: m.a.g.vanderheyden@umcutrecht.nl; Tel.: +31-887558901

† These authors contributed equally to this work.

Received: 28 August 2019; Accepted: 23 October 2019; Published: 25 October 2019

Abstract: The ubiquitously expressed family of inward rectifier potassium (K_{IR}) channels, encoded by *KCNJ* genes, is primarily involved in cell excitability and potassium homeostasis. Channel mutations associate with a variety of severe human diseases and syndromes, affecting many organ systems including the central and peripheral neural system, heart, kidney, pancreas, and skeletal muscle. A number of mutations associate with altered ion channel expression at the plasma membrane, which might result from defective channel trafficking. Trafficking involves cellular processes that transport ion channels to and from their place of function. By alignment of all K_{IR} channels, and depicting the trafficking associated mutations, three mutational hotspots were identified. One localized in the transmembrane-domain 1 and immediately adjacent sequences, one was found in the G-loop and Golgi-export domain, and the third one was detected at the immunoglobulin-like domain. Surprisingly, only few mutations were observed in experimentally determined Endoplasmic Reticulum (ER)exit-, export-, or ER-retention motifs. Structural mapping of the trafficking defect causing mutations provided a 3D framework, which indicates that trafficking deficient mutations form clusters. These "mutation clusters" affect trafficking by different mechanisms, including protein stability.

Keywords: inward rectifier channel; trafficking; alignment; mutation; *KCNJ*; K_{IR}; disease; structure

1. Introduction

Seventy years ago, Katz detected the inward rectification phenomenon for the first time [1]. Its unexpected property of conducting larger inward than outward potassium currents at similar deviations from the potassium equilibrium potential was unprecedented at that time. During the following decades, the understanding of inward rectifier channels was established further, stimulated by biophysical analysis and cloning of K_{IR} genes. Inward rectifying channels—unlike voltage-gated potassium channels (K_v) which open in response to alterations in transmembrane electrostatic potential [2,3]—are primarily gated by intracellular substances (e.g., polyamines and Mg^{2+}). Spermine and spermidine—two polyamines for which micromolar concentrations are sufficient to reach physiological effective levels—cause stronger block of the outward current than Mg^{2+}. The underlying molecular mechanism of rectification was first explained by Lopatin in 1994 [3]. Polyamines enter the channel pore from the cytoplasmic side and subsequently interact with six specific residues (i.e., $K_{IR}2.1$ E224, D259, E299, F254, D255 and D172) [4] in the transmembrane pore domain and its cytosolic pore extension. A similar mechanism of pore-blocking is caused by Mg^{2+}, but weaker.

The inward rectifier channel family consists of strong and weak rectifiers. Strong rectifiers, e.g., $K_{IR}2$ and $K_{IR}3$, are often expressed in excitable cells such as neuronal or muscle cells. Their rectifying properties enable cells to conserve K^+ during action potential formation and facilitate K^+ entry upon cell hyperpolarization. In addition, they contribute to repolarization and stabilization of the resting membrane potential. For example, application of 10 μM barium, at that concentration rather specific for $K_{IR}2$ channel inhibition, lengthened the action potential of guinea-pig papillary muscle preparations by 20 ms [5]. In the heart, $K_{IR}2$ is strongly expressed in the ventricles and less in the atrioventricular node (AVN) [6]; $K_{IR}3$ is mainly expressed in the atrium with much lower levels in the ventricle. Weak rectifier channels, e.g., $K_{IR}1$, $K_{IR}4$, and $K_{IR}5$, are mainly associated with potassium homeostasis and often regulate extracellular potassium concentrations to allow functioning of several ion (co)transporters.

K_{IR} channels are encoded by *KCNJ* genes. Various diseases associate with mutations in *KCNJ* genes. The aim of this review is to correlate disease associated mutations causing aberrant inward rectifier channel trafficking with protein domains important for trafficking by means of channel alignment, and finally to put mutational changes in a structural framework.

2. Classification, Structure, and Expression

The K_{IR} family is divided into seven subfamilies ($K_{IR}1$-7) according to their amino-acid homology [4]. The sequence homology is 40% between subfamilies and rises to 70% within some subfamilies. The structural common features of these channels are that the channel pore is formed by a tetramer of subunits, most often homotetramers (Figure 1). Each subunit has two transmembrane domains (M1 and M2) which are separated by a pore-loop that contains the GYG (or GFG) potassium selectivity filter motif located close to the extracellular side of the membrane. Pore-loop stability depends strongly on one negatively and one positively charged residue, E138 and R148 respectively in $K_{IR}2.1$ [7]. There is a relatively short N-terminus linked to M1 and a longer C-terminus linked to M2 which form the characteristic cytoplasmic extended pore domain (CTD). Despite their structural similarities, the K_{IR} subfamilies also display divergent properties, e.g., sensitivity to extracellular Ba^{2+} or the response to regulatory signals.

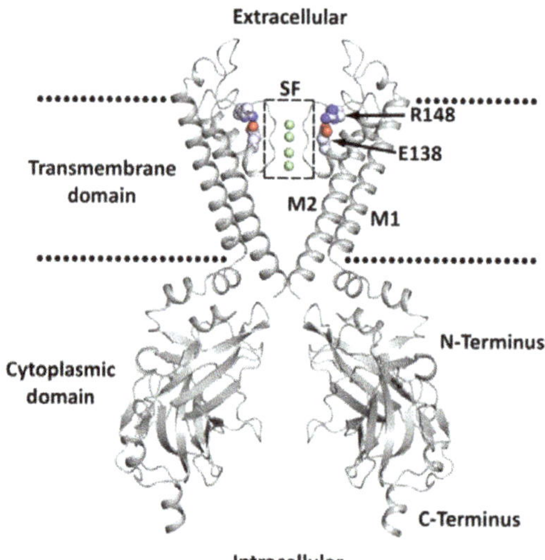

Figure 1. Two opposing domains of the K_{IR} channel with structural common features highlighted. The membrane is indicated by dotted lines. The selectivity filter (SF) is highlighted by a dotted box. Ions inside the SF are shown as green spheres. Residues E138 and R148 ($K_{IR}2.1$) are shown as spheres.

All K_{IR} family members have a widespread expression pattern [4]. Neural tissues strongly express $K_{IR}2$, $K_{IR}3$, $K_{IR}4$, and $K_{IR}5$. Kidneys show high expression of $K_{IR}4$, $K_{IR}5$, and $K_{IR}6$, whereas pancreatic tissue highly expresses $K_{IR}5$ and $K_{IR}6$. The retina shows profound expression of $K_{IR}7$. The heart displays strong expression of $K_{IR}2$, $K_{IR}3$ and $K_{IR}6$. $K_{IR}2$ subfamily-members form the classical I_{K1} current in working ventricular and atrial cardiomyocytes, where they contribute to repolarization and resting membrane potential stability. $K_{IR}3$ (I_{KAch}) members are strongly expressed in the nodal tissues of the heart, where they are involved in heart rate regulation [8]. Further, they are widely expressed in the brain where they have numerous neurological functions [9]. $K_{IR}6$ channels form octamers with the ATP/ADP sensing SUR subunits and couple cellular metabolic status to cardiac repolarization strength.

3. Channel Trafficking

Following their translation in the endoplasmic reticulum (ER), correctly folded K_{IR} channels are transported to the plasma membrane, a process known as forward trafficking, where they exert their main biological role (Figure 2). Upon removal from the plasma membrane, channel proteins can enter the degradation pathway in a process named backward trafficking. In addition, K_{IR} channels can enter several recycling pathways. Each of these processes is well regulated and depends mainly on specific trafficking motifs in the channels primary sequence in concert with specific interacting proteins that direct and/or support subsequent trafficking steps. Incorrectly folded proteins will enter the endoplasmic-reticulum-associated protein degradation pathway.

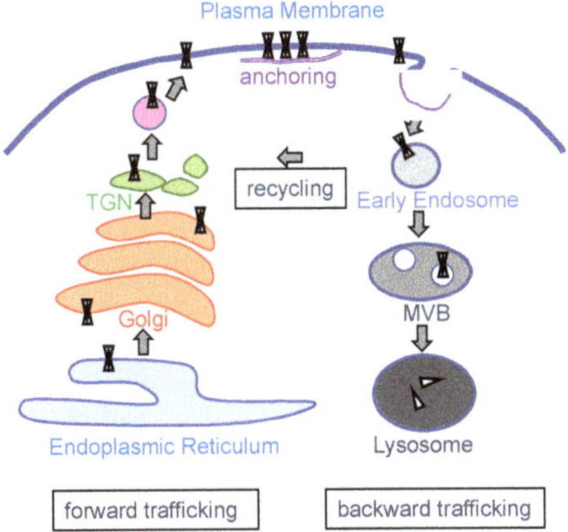

Figure 2. Schematic representation of intracellular trafficking pathways of K_{IR} channels. TGN, trans-Golgi network; MVB, multivesicular body.

ER-export signals have been determined in several K_{IR} channels [10–13], see also Section 5.3, with homology between subfamily members (e.g., FCYENE in $K_{IR}2.x$ channels), although not always among other K_{IR} family members. Not all K_{IR} members possess an ER-export signal, and some might even restrict forward trafficking or stimulate lysosomal breakdown when part of a heteromeric channel, as seen for $K_{IR}3.3$ [12]. Other channels even have ER-retention signals that only become masked upon proper channel assembly, as seen for the $K_{IR}6$ family [13].

Trans-Golgi transport of several K_{IR} channels has been demonstrated to depend on interaction with Golgin tethers that reside in the trans-Golgi network. For example, Golgin-160 interacts with the C-terminal domain of $K_{IR}1.1$ channels which results in increased forward trafficking and an increase in $K_{IR}1.1$ currents [14]. In a similar fashion, Golgin-97 was shown to interact with the C-terminus of $K_{IR}2.1$ and promotes transport to the Golgi-export sites [15]. Golgi-export signals have been characterized in a few K_{IR} channels [16,17]. By a combination of cytoplasmic N- and C-terminal domains, a so-called Golgi-export signal patch is formed that interacts with the AP-1 clathrin adaptor protein.

Protein motifs involved in backward trafficking have been studied less. $K_{IR}1.1$ internalization depends on clathrin-dynamin mediated endocytosis which involved N375 in the $K_{IR}1.1$ putative internalization motif NPN [18]. Internalized $K_{IR}1.1$ channels depend on CORVET and ESCRT protein complexes for subsequent trafficking to the early endosome and the multivesicular body that eventually fuses with the lysosome, respectively [19]. $K_{IR}2.1$ channels are regulated by the ESCRT machinery also [20]. It was demonstrated, by a pharmacological approach, that $K_{IR}2.1$ degradation also depends on clathrin mediated endocytosis and lysosomal activity, and their inhibition resulted in enhanced I_{K1} currents [21,22]. The TPVT motif of the $K_{IR}5.1$ channel protein binds the Nedd4-2 E3 ubiquitin ligase. In $K_{IR}5.1/K_{IR}4.1$ heteromeric complexes, this was suggested to result in ubiquitination and subsequent degradation of $K_{IR}4.1$ in the proteasome [23].

Finally, trafficking, anchoring and plasma membrane localization of K_{IR} channels is regulated by their interaction with scaffolding proteins. The C-terminal $K_{IR}2.2$ SEI PDZ-binding domain interacts with SAP97, PSD-95, Chapsyn-110, SAP102, CASK, Dlg2, Dlg3, Pals2, Veli1, Veli3, Mint1, and abLIM from rat brain lysates, and SAP97, CASK, Veli-3, and Mint1 from rat heart lysates [24–26]. Additionally, interactions between syntrophins, dystrobrevins and the $K_{IR}2.2$ PDZ domain were shown by these authors. $K_{IR}2.1$ and $K_{IR}2.3$ also interact with SAP97 in the heart. Using an NMR approach, it was found that additional residues close to the $K_{IR}2.1$ PDZ domain were involved in PSD-95 interaction [27]. Furthermore, PSD-95 interacts with $K_{IR}4.1$ and $K_{IR}5.1$ in the optic nerve and brain, and PSD-95 interaction is essential for $K_{IR}5.1$ expression at the plasma membrane of HEK293 cells [28,29]. The C-terminal PDZ-binding motif SNV interacts with PSD-95, and $K_{IR}4.1$ mediated current density more than doubled upon PSD-95 cotransfection in HEK293 cells, and increased even threefold upon SAP97 cotransfection [30]. Upon silencing of SAP97, the I_{K1} current decreased due to reduced plasma membrane expression of $K_{IR}2.1$ and $K_{IR}2.2$ ion channels [31]. Residues 307–326 of $K_{IR}2.1$ are involved in interactions with the actin binding protein filamin A. Interestingly, these interactions are unaffected by the Andersen–Tawil deletion $\Delta314/315$ [32]. In arterial smooth muscle cells, filamin A and $K_{IR}2.1$ colocalize in specific regions of the plasma membrane. Although filamin A is not essential for $K_{IR}2.1$ trafficking to the plasma membrane, its absence reduces the amount of $K_{IR}2.1$ channels present at the plasma membrane [32].

Whereas this research field provided many new insights during the last two decades, one has to emphasize that no complete trafficking pathway for any K_{IR} channel protein has been deciphered in detail yet. Furthermore, most of our current knowledge is derived from ectopic expressions systems rather than human native tissue or cells. Currently, we cannot exclude that K_{IR} subtype and/or tissue specific pathways exist. The observations that several diseases associate with K_{IR} channel trafficking malfunction might help us to further understand K_{IR} protein trafficking processes in their natural environments in vivo.

4. Diseases and Syndromes Associated with K_{IR} Channel Dysfunction

A number of human diseases associate with mutations in K_{IR} channels, as indicated in Table 1. Bartter syndrome type II is a salt-losing nephropathy resulting in hypokalemia and alkalosis associated with loss-of-function mutations in $K_{IR}1.1$ channel proteins. $K_{IR}1.1$ channels are essential for luminal extrusion of K^+ in the thick ascending limb of Henle's loop, thereby permitting continued activity of the NKCC2 cotransporter important for sodium resorption [33]. Loss-of-function in $K_{IR}2.1$ causes Andersen–Tawil syndrome characterized by periodic skeletal muscle paralysis, developmental skeletal abnormalities, as well as biventricular tachycardia with or without the presence of long QT. On the other hand, $K_{IR}2.1$ gain-of-function mutations result in cardiac phenotypes, atrial fibrillation and short QT syndrome, explained by increased repolarization capacity and thus shortened cardiac action potentials [34,35]. Thyrotoxic hypokalemic periodic paralysis associated with $K_{IR}2.6$ loss-of-function mutations affect skeletal muscle excitability under thyrotoxic conditions [36]. Keppen–Lubinsky syndrome is an extremely rare condition associated with $K_{IR}3.2$ gain-of-function mutations. Its phenotype encompasses lipodystrophy, hypertonia, hyperreflexia, developmental delay and intellectual disability [37,38]. Familial hyperaldosteronism type III is associated with loss-of-function mutations in $K_{IR}3.4$ channel proteins. The disease is characterized by early onset of severe hypertension and hypokalemia. Mutant $K_{IR}3.4$ channels lack potassium specificity and the resulting inflow of Na^+ and accompanying cell depolarization of zona glomerulosa cells increases intracellular Ca^{2+} concentrations, which activates transcription pathways that raise aldosterone production [39]. Loss-of-function mutations in $K_{IR}3.4$ associate with long QT syndrome 13, which indicates that these acetylcholine activated channels, mostly known from nodal tissues, also play a role in ventricular repolarization processes [40]. EAST (epilepsy, ataxia, sensorineural deafness, tubulopathy)/SeSAME syndrome is a salt-losing nephropathy combined with severe neurological disorders. The disease associated loss-of-function mutations in $K_{IR}4.1$ channels expressed in the distal convoluted tubule, result in hypokalemic metabolic acidosis. Impaired $K_{IR}4.1$ function in glial cells will increase neural tissue potassium levels giving rise to neuron depolarization, whereas reduced potassium concentration in the endolymph affect cochlear hair cell function [41]. Cantú syndrome results from gain-of-function of I_{KATP} channels, either due to mutation in $K_{IR}6.1$ or the SUR2 subunits. Many of these mutations decrease the sensitivity of the channel to ATP-dependent closure [42]. Insulin release by pancreatic beta-cells is regulated by their membrane potential and L-type Calcium channel activity. Depolarization activates Ca^{2+} influx inducing insulin release from intracellular stores into the extracellular fluid. Loss-of-function mutations in $K_{IR}6.2$ result in membrane depolarization and thus insulin release and associate with hyperinsulism and hypoglycemia. Gain-of-function mutations on the other hand impair insulin release and associate with different forms of diabetes [43]. $K_{IR}7.1$ channels are expressed in the apical membrane of retinal pigmented epithelial cells and contribute to K^+ homeostasis in the subretinal space. Loss-of-function mutations in $K_{IR}7.1$ associate with retinal dysfunction observed in Lever congenital amaurosis type 16 and Snowflake vitreoretinal degeneration [44].

In many of the above-mentioned diseases, loss-of-function has been associated with aberrant trafficking, most likely forward trafficking. Nonetheless, enhanced backward trafficking or impaired plasma-membrane anchoring cannot be excluded. However, most gain-of-function mutations are likely not related to trafficking abnormalities. Loss-of-specificity mutations, as seen in some $K_{IR}3.4$ mutations, neither result from trafficking issues.

Table 1. Human diseases associated with abnormal K_{IR} channel function.

Protein	Gene	Syndrome/Disease Character (OMIM)[1]	Main Affected System(s)	Recent Review
$K_{IR}1.1$	*KCNJ1*	Bartter syndrome, type 2 (241200)	Kidney; head; face; ear; eye; vascular; gastrointestinal; skeleton; skeletal muscle; CNS; platelets	[33]
$K_{IR}2.1$	*KCNJ2*	Andersen syndrome (170390) Familial atrium fibrillation 9 (613980) Short QT syndrome 3 (609622)	Head; face; ear; eye; teeth; heart; skeleton; CNS	[34,35]
$K_{IR}2.2$	*KCNJ12*	Non-described		
$K_{IR}2.3$	*KCNJ4*	Non-described		
$K_{IR}2.4$	*KCNJ14*	Non-described		
$K_{IR}2.6$	*KCNJ18*	Thyrotoxic hypokalemic periodic paralysis (613239)	Cardiovascular; skeletal muscle; CNS; eye	[36]
$K_{IR}3.1$	*KCNJ3*	Non-described		
$K_{IR}3.2$	*KCNJ6*	Keppen–Lubinsky Syndrome (614098)	CNS; head; skin; skeleton; eye, face	No review available
$K_{IR}3.3$	*KCNJ9*	Non-described		
$K_{IR}3.4$	*KCNJ5*	Familial hyperaldosteronism 3 (613677) Long QT syndrome 13 (613485)	Cardiovascular; kidney; skeletal muscle	[39,40]
$K_{IR}4.1$	*KCNJ10*	Digenic enlarged vestibular aqueduct (600791) EAST/SESAME syndrome (612780)	Ear (hearing); vascular; kidney; CNS	[41]

Table 1. *Cont.*

Protein	Gene	Syndrome/Disease Character (OMIM)[1]	Main Affected System(s)	Recent Review
$K_{IR}4.2$	*KCNJ15*	Non-described		
$K_{IR}5.1$	*KCNJ16*	Non-described		
$K_{IR}6.1$	*KCNJ8*	Cantú syndrome (239850)	Head; face; cardiovascular; skeleton; hair; CNS	[42]
$K_{IR}6.2$	*KCNJ11*	Transient neonatal diabetes mellitus 3 (610582) Permanent neonatal diabetes with or without neurologic features (606176) Familial hyperinsulinemic hypoglycemia 2 (601820) Maturity-onset diabetes of the young 13 (616329) Susceptible to diabetes mellitus 2 (125853)	Pancreas (beta-cells); CNS	[43]
$K_{IR}7.1$	*KCNJ13*	Leber congenital amaurosis 16 (614186) Snowflake vitreoretinal degeneration (193230)	Eye (retina)	[44]

OMIM[1]: OMIM®—Online Mendelian Inheritance in Man® https://omim.org assessed on 24 July 2019, CNS, central neural system.

5. K_{IR} Protein Alignment of Trafficking Associated Disease Mutations

In order to identify potential protein domains associated with K_{IR} trafficking defects in human disease, we aligned all K_{IR} isoforms and highlighted residues (in red) of which the mutations are experimentally proven to associate with trafficking defects (Figure 3, Supplementary Figure S1). Furthermore, additional mutations in other K_{IR} isoforms at homologues positions, but currently not related to impaired trafficking, are indicated (Supplementary Figure S1, in green). Trafficking associated mutations are found dispersed along the protein sequence, with one "hotspot" in the G-loop and adjacent C-terminal region, and one "hotspot" in and around transmembrane domain 1. Additionally, from a structural point of view, there is another "hotspot" at the immunoglobulin-like domain (IgLD), which is described in Section 6.1.

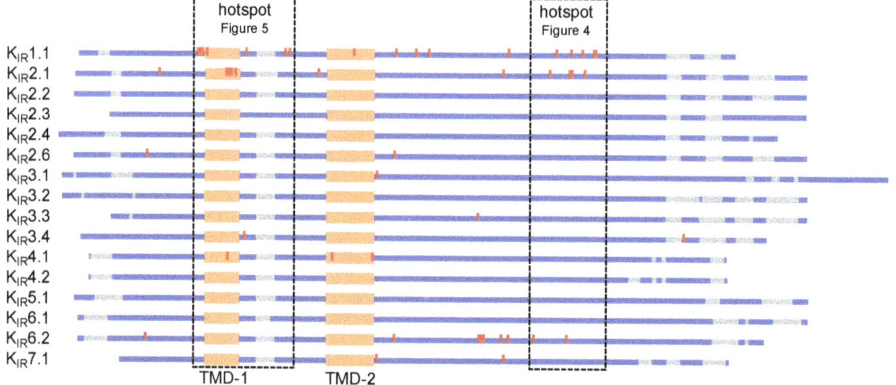

Figure 3. Schematic representation of inward rectifier channels ($K_{IR}1$–7) sequence alignment. Red: mutations associated with aberrant trafficking; mutation hotspots are boxed. Orange: transmembrane domain; blue: K_{IR} protein sequence; gray: sequence gap.

5.1. C-Terminal Trafficking Mutation Hotspot

Figure 4 depicts the alignment of the C-terminal hotspot, having ten trafficking associated mutations/deletions located in $K_{IR}1.1$, $K_{IR}2.1$ and/or $K_{IR}6.2$ over a stretch of 44 residues. This region also covers the filamin A interaction domain of $K_{IR}2.1$ (307–326) [32]. The loss-of-function E282K mutation in $K_{IR}6.2$ is associated with congenital hyperinsulinism [45]. This mutation affects normal function of a highly conserved di-acidic ER exit signal (DxE) that prevents mutant channels to enter ER exit sites, which thus fail to traffic to the plasma membrane [45]. The Andersen–Tawil loss-of-function $K_{IR}2.1$ mutation V302M is located in the G-loop and displays intracellular, but no plasma-membrane expression, upon transfection of HEK293 cells [46]. However, Ma et al., [47] demonstrated that the $K_{IR}2.1$ V302M mutation does not affect trafficking. The Bartter syndrome associated loss-of-function A306T mutation in $K_{IR}1.1$ is also located in the G-loop. Its expression in the *Xenopus* oocyte membrane is strongly reduced compared to wildtype channels [48]. At homologues positions, disease causing mutations have been found in $K_{IR}2.1$ [49] and $K_{IR}6.2$ [50] whose cause for loss- and gain-of-function, respectively, is unknown but may be caused by trafficking abnormalities. Hyperinsulinism associated $K_{IR}6.2$ R301H/G/C/P loss-of-function mutations are located just C-terminal from the G-loop domain. These mutants display reduced surface expression when expressed in COSm6 or INS cells [51]. Interestingly, however, the R301A mutation displays normal expression at the plasma membrane [51]. At homologues positions, mutations have also been found in $K_{IR}1.1$ [52] and $K_{IR}4.1$ [53].

```
KIR1.1   QQDFELVVFLDGTVESTSATCQVRTSYVPEEVLWGYRFAPIVSK   331
KIR2.1   NADFEIVVILEGMVEATAMTTQCRSSYLANEILWGHRYEPVLFE   332
KIR2.2   TDDFEIVVILEGMVEATAMTTQARSSYLANEILWGHRFEPVLFE   333
KIR2.3   SEDFEIVVILEGMVEATAMTTQARSSYLASEILWGHRFEPVVFE   324
KIR2.4   RADFELVVILEGMVEATAMTTQCRSSYLPGELLWGHRFEPVLFQ   337
KIR2.6   TDDFEIVVILEGMVEATAMTTQARSSYLANEILWGHRFEPVLFE   333
KIR3.1   TEQFEIVVILEGIVETTGMTCQARTSYTEDEVLWGHRFFPVISL   333
KIR3.2   KEELEIVVILEGMVEATGMTCQARSSYITSEILWGYRFTPVLTL   342
KIR3.3   RDDFEIVVILEGMVEATGMTCQARSSYLVDEVLWGHRFTSVLTL   310
KIR3.4   QEEFEVVVILEGMVEATGMTCQARSSYMDTEVLWGHRFTPVLTL   339
KIR4.1   EGDFELVLILSGTVESTSATCQVRTSYLPEEILWGYEFTPAISL   317
KIR4.2   EKEFELVVLLNATVESTSAVCQSRTSYIPEEIYWGFEFVPVVSL   317
KIR5.1   KDNFEILVTFIYTGDSTGTSHQSRSSYVPREILWGHRFNDVLEV   316
KIR6.1   NQDLEVIVILEGVVETTGITTQARTSYIAEEIQWGHRFVSIVTE   330
KIR6.2   HQDLEIIVILEGVVETTGITTQARTSYLADEILWGQRFVPIVAE   321
KIR7.1   PSHFELVVFLSAMQEGTGEICQRRTSYLPSEIMLHHCFASLLTR   305
         [DE-E]        [-G-LOOP--]    SY---EI-W
         di-acidic                   Golgi-export
         ER exit
```

Figure 4. $K_{IR}1$–7 sequence alignment of the C-terminal mutation hotspot. Red: Mutations associated with aberrant trafficking; Green: Residues whose mutations are currently not related to impaired trafficking. Numbers at the right refer to the last amino-acid residue in the respective sequence shown. Conserved residues among all K_{IR} members are shaded gray. Locations of the di-acidic ER exit, G-loop and the Golgi-Export signal sequence (see text) are indicated below the alignment.

By use of homology comparison and structure guided mutagenesis, a common Golgi-export signal patch was found to be formed by a C-terminal stretch of hydrophobic residues and basic residues from the N-terminus [16,17]. The C-terminal stretch sequence is formed by residues SYxxxEIxW indicated in Figure 4. Two Bartter syndrome associated $K_{IR}1.1$ (Y314C; L320P) and two Andersen–Tawil syndrome associated $K_{IR}2.1$ (delSY; W322S) confirmed trafficking mutations have been described in this region [46,48,54]. Interestingly, the W residue is not conserved in the $K_{IR}7.1$ channel protein, which may indicate K_{IR} subtype specific use of the entire Golgi-export signal motif. Two additional $K_{IR}1.1$ mutations leading to altered trafficking, R324L and F325C have been located directly C-terminal from the Golgi-export signal stretch [48,52].

Fallen et al. [55] showed that mutations A198T and Y314C in the IgLD, located in the CTD of $K_{IR}1.1$, are associated with defects in channel trafficking and gating, see also Section 6.1. Y314C is present within the C-terminal trafficking mutation hotspot and part of the Golgi-export signal as discussed above. If incorrectly folded, the aberrant $K_{IR}1.1$ proteins will enter the endoplasmic-reticulum-associated protein degradation pathway [56].

5.2. Transmembrane Region 1 Mutation Hotspot

Figure 5 depicts the alignment of transmembrane domain 1 and adjacent sequences, containing 15 trafficking associated mutations located in $K_{IR}1.1$, $K_{IR}2.1$, $K_{IR}3.4$ or $K_{IR}4.1$. Three mutations in the cytoplasmic domain, positioned just in front of the first transmembrane region in $K_{IR}1.1$ (T71M, V72M and D74Y), have been demonstrated to strongly decrease plasma-membrane expression and mutant channels were retained in the cytoplasm [48,57]. However, membrane expression of T71M in *Xenopus* oocytes could be rescued by increasing the amount of injected RNA, in contrast to the other two mutations. For $K_{IR}1.1$ T71M, mutations at the homologues positions were found in $K_{IR}2.1$ (T75) [58–62] and $K_{IR}6.2$ (T62) [63,64] associated with Andersen–Tawil syndrome and Familial hyperinsulinemic hypoglycemia type 2, respectively. The $K_{IR}2.1$ T75R protein was observed at the plasma membrane upon overexpression in HL1 cells [60]. Moreover, T75A, T75R and T75M channel proteins were also expressed at the plasma membrane in *Xenopus* oocytes, HEK293 or COS-7 cells [58,61,65]. In contrast, impaired plasma-membrane localization of T75M $K_{IR}2.1$ was observed in HEK293 cells in another study [62]. Two $K_{IR}2.1$ mutations, i.e., D78G and D78Y, are at the equivalent position as D74 in $K_{IR}1.1$, and also D78Y was found at the plasma membrane in *Xenopus* oocytes and HEK293 cells [59,65,66]. These comparisons indicate that findings on plasma-membrane expression may be influenced by the degree of overexpression and cell type. $K_{IR}2.1$ T75 and D78 residues are positioned on the hydrophilic side of the slide helix that interacts with the cytoplasmic domain. The D78Y mutation disrupts this interaction [65].

```
KIR1.1   TTVLDLKWRYKMTIFITAFLGSWFFFGLLWYAVAYIHKDLPE--FHP-------------SANHTPCVENI   125
KIR2.1   TTCVDIRWRWMLVIFCLAFVLSWLFFGCVFWLIALLHGDLDASK----------------EGKACVSEV    126
KIR2.2   TTCVDIRWRYMLLIFSLAFLASWLLFGIIFWVIAVAHGDLEPAEGR--------------GRTPCVMQV    127
KIR2.3   TTCVDTRWRYMLMIFSAAFLVSWLFFGLLFWCIAFFHGDLEASPGVPAAGGPAAGGGGAAPVAPKPCIMHV   118
KIR2.4   TTCVDVRWRWMCLLFSCSFLASWLLFGLAFWLIASLHGDLAAPP----------------PPAPCFSHV   131
KIR2.6   TTCVDIRWRYMLLIFSLAFLASWLLFGVIFWVIAVAHGDLEPAEGH--------------GRTPCVMQV    127
KIR3.1   TTLVDLKWRWNLFIFILTYTVAWLFMASMWWVIAYTRGDLNKAHVG--------------NYTPCVANV    127
KIR3.2   TTLVDLKWRFNLLIFVMVYTVTWLFFGMIWWLIAYIRGDMDHIEDP--------------SWTPCVTNL    136
KIR3.3   TTLVDLQWRLSLLFFVLAYALTWLFFGAIWWLIAYGRGDLEHLEDT--------------AWTPCVNNL    104
KIR3.4   TTLVDLKWRFNLLVFTMVYTVTWLFFGFIWWLIAYIRGDLDHVGDQ--------------EWIPCVENL    133
KIR4.1   TTFIDMQWRYKLLLFSATFAGTWFLFGVVWYLVAVAHGDLLE--LDP------------PANHTPCVVQV   112
KIR4.2   TTVIDMKWRYKLTLFAATFVMTWFLFGVIYYAIAFIHGDLEP--GEP------------ISNHTPCIMKV   111
KIR5.1   TTLVDTKWRHMFVIFSLSYILSWLIFGSVFWLIAFHHGDLLNDP----------------DITPCVDNV   115
KIR6.1   TTLVDLKWRHTLVIFTMSFLCSWLLFAIMWWLVAFAHGDIYAYMEKS-------GMEKSGLESTVCVTNV   124
KIR6.2   TTLVDLKWPHTLLIFTMSFLCSWLLFAMAWWLIAFAHGDLAPS--------------EGTAEPCVTSI    114
KIR7.1   GILMDMRWRWMMLVFSASFVVHWLVFAVLWYVLAEMNGDLELDHDAP-------------PENHTICVKYI   103
                  [-transmembrane-domain 1--]
```

Figure 5. $K_{IR}1$–7 sequence alignment of the transmembrane domain 1 mutation hotspot. Red: mutations associated with aberrant trafficking; Green: residues whose mutations are currently not related to impaired trafficking. Numbers at the right refer to the last amino-acid residue in the respective sequence shown. Conserved residues among all K_{IR} members are shaded gray. Location of the transmembrane domain 1 (orange) is indicated below the alignment.

Trafficking associated mutations in the highly conserved transmembrane region 1 are described for $K_{IR}1.1$, $K_{IR}2.1$ and $K_{IR}4.1$ [46,48,67–69]. Expression of $K_{IR}1.1$ Y79H in the *Xenopus* oocyte plasma-membrane increases upon increasing the amount of RNA injection by ten-fold [48]. The $K_{IR}2.1$ L94P, Δ95-98 and $K_{IR}4.1$ G77R channel proteins localize intracellularly [46,68,69]. The molecular mechanisms by which these mutations affect normal trafficking remain to be solved. However, interactions with wildtype subunits appear not to be affected and may explain the dominant negative properties of these mutations. The familial sinus node disease associated $K_{IR}3.4$ W101C gain-of-function mutation is located at a position homologues to $K_{IR}2.1$ W96 [70]. In an ectopic expression system, the $K_{IR}3.4$ W101C protein is expressed at the plasma membrane, however it decreased surface expression of $K_{IR}3.1$ when co-expressed [70].

Confirmed trafficking associated mutations C-terminal from the transmembrane region 1 are found in $K_{IR}1.1$ and $K_{IR}3.4$ [48,71,72]. $K_{IR}1.1$ D108H and V122E mutants did not display membrane staining in *Xenopus* oocytes or HEK293 cells [48]. When comparing single channel characteristics with macroscopic currents, it was concluded that loss-of-function of $K_{IR}1.1$ N124K was caused by a reduction of functional channels at the plasma membrane [71]. The $K_{IR}3.4$ R115W mutation was obtained from aldosterone-producing adenoma linked to hyperaldosteronism, and displayed decreased plasma-membrane expression in HEK293 cells [72]. A mutation, at a position homologues to V122 in $K_{IR}1.1$, has also been identified in $K_{IR}2.1$ [59].

5.3. N-Terminal Golgi-Export Patch, $K_{IR}2.x$ ER Export, and $K_{IR}6.x$ ER Exit and Retention Signals

As indicated above, the so-called Golgi-export patch consists of interaction of a C-terminal and N-terminal domain [16,17]. Mutations in the C-terminal domain have been found (see Section 5.1). However, only few mutations have been described in the N-terminal part ($K_{IR}2.1$ G52V; $K_{IR}2.6$ R43C) that result in reduced plasma membrane expression by hampering Golgi export [73,74]. Thus far, no mutations of residue R20 in $K_{IR}2.3$, which is required for Golgi export [17], or at the homologues position in any other K_{IR} channel protein have been identified.

$K_{IR}2.x$ channels share a short C-terminal ER-export signal (FCYENE) [10,11]. $K_{IR}3.2$ contains N-terminal (DQDVESPV) and C-terminal (ELETEEEE) ER-export signals, whereas $K_{IR}3.4$ possesses the N-terminal NQDMEIGV ER-export signal [12]. Remarkable, we did not encounter any trafficking associated mutations in any of these domains. In contrast, one mutation (E282K) was present in the di-acidic ER exit signal of $K_{IR}6.2$ as discussed in Section 5.1. $K_{IR}6.x$ and SURx channel proteins contain a C-terminal ER-retention signal (RKR) [13]. Upon channel assembly, retention signals from both proteins are shielded, supporting subsequent ER-export. No trafficking associated mutations were found in these retention signals in $K_{IR}6.x$ channel proteins.

We therefore propose that mutations in Golgi-export domains have more severe clinical implications than mutations in ER-export/retention signals.

6. Structural Mapping of Trafficking Defect Causing Mutants

Disease causing mutations associated with trafficking deficiencies were mapped onto the common structural scaffold of a recently published high resolution $K_{IR}2.2$ structure [75]. As illustrated in Figure 6, mutations are globally distributed.

A group of mutations clusters at regions important for channel gating, including the PIP_2 binding site (T71M, V72E, D74Y and Y79H in $K_{IR}1.1$), the helix bundle crossing gate (A167V in $K_{IR}4.1$; R162W/Q in $K_{IR}7.1$) [76–78], as well as the G-loop gate (V302M, $K_{IR}2.1$). It can be expected that these mutations have strong effects on the conformational equilibrium, thereby impairing normal protein function. It is likely that these mutants lead to structurally less-stable proteins, thereby making them more susceptible for degradation. Interestingly, 58% of the currently known trafficking defect causing mutations in K_{IR} channel proteins cluster in the cytoplasmic domain, which has been shown to be crucial for efficient folding in K_VAP channels [79].

Another cluster of mutants (D108H, V122E and N124K in $K_{IR}1.1$; R115W in $K_{IR}3.4$; C140R in $K_{IR}4.1$) is found on surface exposed loops of the channel. Except for C140R in $K_{IR}4.1$ [68], which is part of a disulfide bridge [80], none of the mutations causes changes in polarity or is at important structural motifs, leaving it unclear why these mutants cause trafficking defects.

Mutations G77R in $K_{IR}4.1$ [68] and C101R in $K_{IR}2.1$, located on the membrane facing side and near the center of transmembrane helix M1, cause changes in the helical properties and hydrophobicity. It is thus conceivable that they severely affect helical stability and possibly membrane insertion. It has been shown in numerous studies that the cost for exposing arginine to lipid hydrocarbons is prohibitively high [81]. Interestingly, none of the identified disease mutations is located at the interface between subunits.

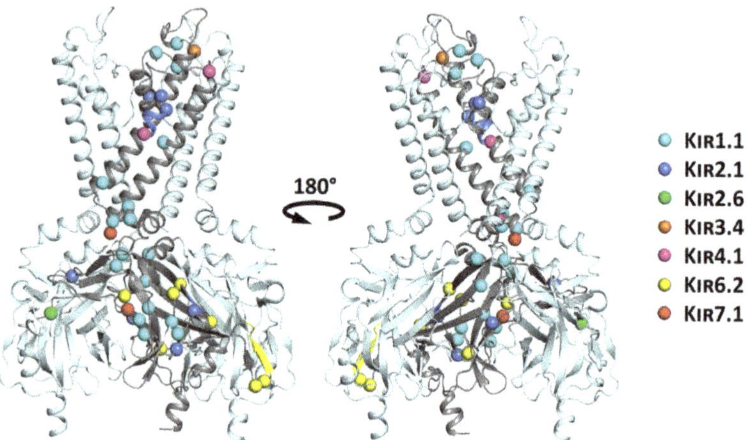

Figure 6. Structural mapping of trafficking mutants mapped on the $K_{IR}2.2$ structure. For clarity reasons, only three of the four subunits are shown in side view, with the disease associated mutations highlighted in one subunit only. Mutations of different K_{IR} channel family members are color-coded and shown as spheres of their respective Cα atoms. Deletions are indicated by colored regions on the secondary structure elements.

6.1. Structure-Based Hotspot in the IgLD Beta Barrel of the CTD

Two antenatal Bartter syndrome loss-of-function mutations A198T and Y314C, located in the IgLD have been shown to impair forward trafficking and gating of $K_{IR}1.1$ channels, possibly influencing the core stability of this domain [55]. Interestingly, a trafficking affecting mutation in a homologous position to $K_{IR}1.1$ A198 has been identified in $K_{IR}2.6$ (A200P) [74], and another mutation thus far not associated with trafficking has been identified in $K_{IR}6.2$ (A187V). As shown in Figure 7, quite a large number of disease causing mutations, including A306T (implicated in trafficking) [48], R311W and L320P in $K_{IR}1.1$ (no data on trafficking), S314-Y315 deletion in $K_{IR}2.1$ (implicated in trafficking) [16], E282K (prevents ER-export and surface expression of the channel) [45] or L241P in $K_{IR}7.1$ (implicated in trafficking) [78], have been reported in the literature. This, as well as previous work [55], suggests that this structural motif might be a crucial hotspot implicated in trafficking of K_{IR} channels.

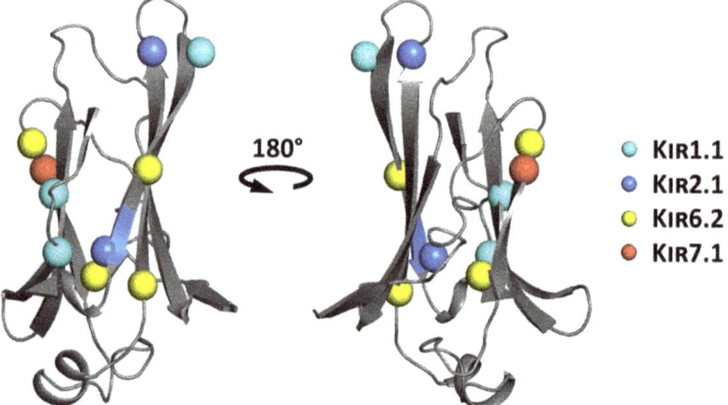

Figure 7. Structure-based IgLD hotspot (mapped on the $K_{IR}2.2$ structure), with disease associated mutations highlighted. Mutations of the different family members are color-coded and shown as spheres of their respective Cα atoms.

7. Conclusions

Mutations in K_{IR} potassium ion channels associate with a variety of human diseases in which electrophysiological and potassium homeostasis aberrations are explaining etiology. Many of the mutations associate with abnormal, mostly decreased forward, ion channel trafficking. Trafficking associated mutations are present throughout the primary sequence, but they concentrate in cytoplasmic domains in which channel structures involved in Golgi-export are clinically more important than ER-export regions. Another group of mutations are found in regions important for gating and most likely affect protein folding and stability. Therefore, mutation associated K_{IR} trafficking defects are likely caused by 1) defective interaction with the trafficking machinery due to mutations in specific trafficking motifs, and 2) channel misfolding, destabilization and subsequent endoplasmic-reticulum-associated protein degradation due to mutations in residues important for channel structure.

Supplementary Materials: The following are available online at http://www.mdpi.com/2218-273X/9/11/650/s1, Figure S1: K_{IR}1-7 sequence alignment.

Author Contributions: M.A.G.v.d.H. and A.S.-W. conceptualized the idea. E.-M.Z.-P., M.Q., M.B., A.S.-W. and M.A.G.v.d.H. wrote, edited and reviewed the manuscript. E.-M.Z.-P. and M.A.G.v.d.H. prepared the figures. M.A.G.v.d.H. coordinated the writing up and the submission process. E.-M.Z.-P., M.Q., M.B., A.S.-W. and M.A.G.v.d.H. approved the final version for submission. Funding acquisition by M.Q. and M.A.G.v.d.H.

Funding: Muge Qile was funded by Chinese Scholarship Council.

Conflicts of Interest: The authors declare no conflicts of interest. The funders had no role in the design of the study; in the collection, analyses, or interpretation of data; in the writing of the manuscript, or in the decision to publish the results.

References

1. Katz, B. Les constantes électriques de la membrane du muscle. *Arch. Sci. Physiol.* **1949**, *2*, 285–299.

2. Matsuda, H.; Saigusa, A.; Irisawa, H. Ohmic conductance through the inwardly rectifying K channel and blocking by internal Mg^{2+}. *Nature* **1987**, *325*, 156–159. [CrossRef] [PubMed]

3. Lopatin, A.N.; Makhina, E.N.; Nichols, C.G. Potassium channel block by cytoplasmic polyamines as the mechanism of intrinsic rectification. *Nature* **1994**, *372*, 366–369. [CrossRef] [PubMed]

4. De Boer, T.P.; Houtman, M.J.; Compier, M.; Van der Heyden, M.A. The mammalian K_{IR}2. x inward rectifier ion channel family: Expression pattern and pathophysiology. *Acta Physiol.* **2010**, *199*, 243–256. [CrossRef]

5. Wang, L.; Chiamvimonvat, N.; Duff, H.J. Interaction between selected sodium and potassium channel blockers in guinea pig papillary muscle. *J. Pharmacol. Exp. Ther.* **1993**, *264*, 1056–1062.

6. Kokubun, S.; Nishimura, M.; Noma, A.; Irisawa, H. Membrane currents in the rabbit atrioventricular node cell. *Pflügers Arch.* **1982**, *393*, 15–22. [CrossRef]

7. Yang, J.; Yu, M.; Jan, Y.N.; Jan, L.Y. Stabilization of ion selectivity filter by pore loop ion pairs in an inwardly rectifying potassium channel. *Proc. Natl. Acad. Sci. USA* **1997**, *94*, 1568–1572. [CrossRef]

8. Krapivinsky, G.; Gordon, E.A.; Wickman, K.; Velimirović, B.; Krapivinsky, L.; Clapham, D.E. The G-protein-gated atrial K^+ channel I_{KACh} is a heteromultimer of two inwardly rectifying K^+-channel proteins. *Nature* **1995**, *374*, 135–141. [CrossRef]

9. Lüscher, C.; Slesinger, P.A. Emerging roles for G protein-gated inwardly rectifying potassium (GIRK) channels in health and disease. *Nat. Rev. Neurosci.* **2010**, *11*, 301–315. [CrossRef]

10. Ma, D.; Zerangue, N.; Lin, Y.F.; Collins, A.; Yu, M.; Jan, Y.N.; Jan, L.Y. Role of ER export signals in controlling surface potassium channel numbers. *Science* **2001**, *291*, 316–319. [CrossRef]

11. Stockklausner, C.; Ludwig, J.; Ruppersberg, J.P.; Klöcker, N. A sequence motif responsible for ER export and surface expression of Kir2.0 inward rectifier K^+ channels. *FEBS Lett.* **2001**, *493*, 129–133. [CrossRef]

12. Ma, D.; Zerangue, N.; Raab-Graham, K.; Fried, S.R.; Jan, Y.N.; Jan, L.Y. Diverse trafficking patterns due to multiple traffic motifs in G protein-activated inwardly rectifying potassium channels from brain and heart. *Neuron* **2002**, *33*, 715–729. [CrossRef]

13. Zerangue, N.; Schwappach, B.; Jan, Y.N.; Jan, L.Y. A new ER trafficking signal regulates the subunit stoichiometry of plasma membrane K(ATP) channels. *Neuron* **1999**, *22*, 537–548. [CrossRef]

14. Bundis, F.; Neagoe, I.; Schwappach, B.; Steinmeyer, K. Involvement of Golgin-160 in cell surface transport of renal ROMK channel: Co-expression of Golgin-160 increases ROMK currents. *Cell Physiol. Biochem.* **2006**, *17*, 1–12. [CrossRef]

15. Taneja, T.K.; Ma, D.; Kim, B.Y.; Welling, P.A. Golgin-97 Targets Ectopically Expressed Inward Rectifying Potassium Channel, Kir2.1, to the trans-Golgi Network in COS-7 Cells. *Front. Physiol.* **2018**, *9*, 1070. [CrossRef]

16. Ma, D.; Taneja, T.K.; Hagen, B.M.; Kim, B.Y.; Ortega, B.; Lederer, W.J.; Welling, P.A. Golgi export of the Kir2.1 channel is driven by a trafficking signal located within its tertiary structure. *Cell* **2011**, *145*, 1102–1115. [CrossRef]

17. Li, X.; Ortega, B.; Kim, B.; Welling, P.A. A Common Signal Patch Drives AP-1 Protein-dependent Golgi Export of Inwardly Rectifying Potassium Channels. *J. Biol. Chem.* **2016**, *291*, 14963–14972. [CrossRef]

18. Zeng, W.Z.; Babich, V.; Ortega, B.; Quigley, R.; White, S.J.; Welling, P.A.; Huang, C.L. Evidence for endocytosis of ROMK potassium channel via clathrin-coated vesicles. *Am. J. Physiol. Renal Physiol.* **2002**, *283*, 630–639. [CrossRef]

19. Mackie, T.D.; Kim, B.Y.; Subramanya, A.R.; Bain, D.J.; O'Donnell, A.F.; Welling, P.A.; Brodsky, J.L. The endosomal trafficking factors CORVET and ESCRT suppress plasma membrane residence of the renal outer medullary potassium channel (ROMK). *J. Biol. Chem.* **2018**, *293*, 3201–3217. [CrossRef]

20. Kolb, A.R.; Needham, P.G.; Rothenberg, C.; Guerriero, C.J.; Welling, P.A.; Brodsky, JL. ESCRT regulates surface expression of the Kir2.1 potassium channel. *Mol. Biol. Cell* **2014**, *25*, 276–289. [CrossRef]

21. Jansen, J.A.; de Boer, T.P.; Wolswinkel, R.; van Veen, T.A.; Vos, M.A.; van Rijen, H.V.M.; van der Heyden, M.A.G. Lysosome mediated Kir2.1 breakdown directly influences inward rectifier current density. *Biochem. Biophys. Res. Commun.* **2008**, *367*, 687–692. [CrossRef] [PubMed]

22. Varkevisser, R.; Houtman, M.J.; Waasdorp, M.; Man, J.C.; Heukers, R.; Takanari, H.; Tieland, R.G.; van Bergen En Henegouwen, P.M.; Vos, M.A.; van der Heyden, M.A. Inhibiting the clathrin-mediated endocytosis pathway rescues K(IR)2.1 downregulation by pentamidine. *Pflugers Arch.* **2013**, *465*, 247–259. [CrossRef] [PubMed]

23. Wang, M.X.; Su, X.T.; Wu, P.; Gao, Z.X.; Wang, W.H.; Staub, O.; Lin, D.H. Kir5.1 regulates Nedd4-2-mediated ubiquitination of Kir4.1 in distal nephron. *Am. J. Physiol. Renal Physiol.* **2018**, *315*, F986–F996. [CrossRef] [PubMed]

24. Leonoudakis, D.; Mailliard, W.; Wingerd, K.; Clegg, D.; Vandenberg, C. Inward rectifier potassium channel Kir2.2 is associated with synapse-associated protein SAP97. *J. Cell Sci.* **2001**, *114*, 987–998. [PubMed]

25. Leonoudakis, D.; Conti, L.R.; Radeke, C.M.; McGuire, L.M.; Vandenberg, C.A. A multiprotein trafficking complex composed of SAP97, CASK, Veli, and Mint1 is associated with inward rectifier Kir2 potassium channels. *J. Biol. Chem.* **2004**, *279*, 19051–19063. [CrossRef] [PubMed]

26. Leonoudakis, D.; Conti, L.R.; Anderson, S.; Radeke, C.M.; McGuire, L.M.; Adams, M.E.; Froehner, S.C.; Yates, J.R., 3rd; Vandenberg, C.A. Protein trafficking and anchoring complexes revealed by proteomic analysis of inward rectifier potassium channel (Kir2.x)-associated proteins. *J. Biol. Chem.* **2004**, *279*, 22331–22346. [CrossRef]

27. Pegan, S.; Tan, J.; Huang, A.; Slesinger, P.A.; Riek, R.; Choe, S. NMR studies of interactions between C-terminal tail of Kir2.1 channel and PDZ1,2 domains of PSD95. *Biochemistry* **2007**, *46*, 5315–5322. [CrossRef]

28. Brasko, C.; Hawkins, V.; De La Rocha, I.C.; Butt, A.M. Expression of Kir4.1 and Kir5.1 inwardly rectifying potassium channels in oligodendrocytes, the myelinating cells of the CNS. *Brain Struct. Funct.* **2017**, *222*, 41–59. [CrossRef]

29. Tanemoto, M.; Fujita, A.; Higashi, K.; Kurachi, Y. PSD-95 mediates formation of a functional homomeric Kir5.1 channel in the brain. *Neuron* **2002**, *34*, 387–397. [CrossRef]

30. Horio, Y.; Hibino, H.; Inanobe, A.; Yamada, M.; Ishii, M.; Tada, Y.; Satoh, E.; Hata, Y.; Takai, Y.; Kurachi, Y. Clustering and enhanced activity of an inwardly rectifying potassium channel, Kir4.1, by an anchoring protein, PSD-95/SAP90. *J. Biol. Chem.* **1997**, *272*, 12885–12888. [CrossRef]

31. Vaidyanathan, R.; Taffet, S.M.; Vikstrom, K.L.; Anumonwo, J.M. Regulation of cardiac inward rectifier potassium current (I(K1)) by synapse-associated protein-97. *J. Biol. Chem.* **2010**, *285*, 28000–28009. [CrossRef] [PubMed]

32. Sampson, L.J.; Leyland, M.L.; Dart, C. Direct interaction between the actin-binding protein filamin-A and the inwardly rectifying potassium channel, Kir2.1. *J. Biol. Chem.* **2003**, *278*, 41988–41997. [CrossRef] [PubMed]

33. Seyberth, H.W.; Weber, S.; Kömhoff, M. Bartter's and Gitelman's syndrome. *Curr. Opin. Pediatr.* **2017**, *29*, 179–186. [CrossRef] [PubMed]

34. Nguyen, H.L.; Pieper, G.H.; Wilders, R. Andersen-Tawil syndrome: Clinical and molecular aspects. *Int. J. Cardiol.* **2013**, *170*, 1–16. [CrossRef] [PubMed]

35. Hancox, J.C.; Whittaker, D.G.; Du, C.; Stuart, A.G.; Zhang, H. Emerging therapeutic targets in the short QT syndrome. *Expert Opin. Ther. Targets* **2018**, *22*, 439–451. [CrossRef] [PubMed]

36. Fialho, D.; Robert, C.G.; Emma, M. Periodic paralysis. In *Handbook of Clinical Neurology*; Elsevier: Amsterdam, The Netherlands, 2018; Volume 148, pp. 505–520.

37. Masotti, A.; Uva, P.; Davis-Keppen, L.; Basel-Vanagaite, L.; Cohen, L.; Pisaneschi, E.; Celluzzi, A.; Bencivenga, P.; Fang, M.; Tian, M.; et al. Keppen-Lubinsky syndrome is caused by mutations in the inwardly rectifying K+ channel encoded by KCNJ6. *Am. J. Hum. Genet.* **2015**, *96*, 295–300. [CrossRef]

38. Horvath, G.A.; Zhao, Y.; Tarailo-Graovac, M.; Boelman, C.; Gill, H.; Shyr, C.; Lee, J.; Blydt-Hansen, I.; Drögemöller, B.I.; Moreland, J.; et al. Gain-of-function KCNJ6 Mutation in a Severe Hyperkinetic Movement Disorder Phenotype. *Neuroscience* **2018**, *384*, 152–164. [CrossRef]

39. Korah, H.E.; Scholl, U.I. An Update on Familial Hyperaldosteronism. *Horm. Metab. Res.* **2015**, *47*, 941–946. [CrossRef]

40. Bohnen, M.S.; Peng, G.; Robey, S.H.; Terrenoire, C.; Iyer, V.; Sampson, K.J.; Kass, R.S. Molecular Pathophysiology of Congenital Long QT Syndrome. *Physiol. Rev.* **2017**, *97*, 89–134. [CrossRef]

41. Abdelhadi, O.; Iancu, D.; Stanescu, H.; Kleta, R.; Bockenhauer, D. EAST syndrome: Clinical, pathophysiological, and genetic aspects of mutations in KCNJ10. *Rare Dis.* **2016**, *4*, e1195043. [CrossRef]

42. Nichols, C.G.; Singh, G.K.; Grange, D.K. KATP channels and cardiovascular disease: Suddenly a syndrome. *Circ. Res.* **2013**, *112*, 1059–1072. [CrossRef] [PubMed]

43. Tinker, A.; Aziz, Q.; Li, Y.; Specterman, M. ATP-Sensitive Potassium Channels and Their Physiological and Pathophysiological Roles. *Compr. Physiol.* **2018**, *8*, 1463–1511. [PubMed]

44. Kumar, M.; Pattnaik, B.R. Focus on Kir7.1: Physiology and channelopathy. *Channels* **2014**, *8*, 488–495. [CrossRef] [PubMed]

45. Taneja, T.K.; Mankouri, J.; Karnik, R.; Kannan, S.; Smith, A.J.; Munsey, T.; Christesen, H.B.; Beech, D.J.; Sivaprasadarao, A. Sar1-GTPase-dependent ER exit of KATP channels revealed by a mutation causing congenital hyperinsulinism. *Hum. Mol. Genet.* **2009**, *18*, 2400–2413. [CrossRef]

46. Bendahhou, S.; Donaldson, M.R.; Plaster, N.M.; Tristani-Firouzi, M.; Fu, Y.H.; Ptácek, L.J. Defective potassium channel Kir2.1 trafficking underlies Andersen-Tawil syndrome. *J. Biol. Chem.* **2003**, *278*, 51779–51785. [CrossRef]

47. Ma, D.; Tang, X.D.; Rogers, T.B.; Welling, P.A. An andersen-Tawil syndrome mutation in Kir2.1 (V302M) alters the G-loop cytoplasmic K$^+$ conduction pathway. *J. Biol. Chem.* **2007**, *282*, 5781–5789. [CrossRef]

48. Peters, M.; Ermert, S.; Jeck, N.; Derst, C.; Pechmann, U.; Weber, S.; Schlingmann, K.P.; Seyberth, H.W.; Waldegger, S.; Konrad, M. Classification and rescue of ROMK mutations underlying hyperprostaglandin E syndrome/antenatal Bartter syndrome. *Kidney Int.* **2003**, *64*, 923–932. [CrossRef]

49. Choi, B.O.; Kim, J.; Suh, B.C.; Yu, J.S.; Sunwoo, I.N.; Kim, S.J.; Kim, G.H.; Chung, K.W. Mutations of KCNJ2 gene associated with Andersen-Tawil syndrome in Korean families. *J. Hum. Genet.* **2007**, *52*, 280–283. [CrossRef]

50. Gloyn, A.L.; Pearson, E.R.; Antcliff, J.F.; Proks, P.; Bruining, G.J.; Slingerland, A.S.; Howard, N.; Srinivasan, S.; Silva, J.M.; Molnes, J.; et al. Activating mutations in the gene encoding the ATP-sensitive potassium-channel subunit Kir6.2 and permanent neonatal diabetes. *N. Engl. J. Med.* **2004**, *350*, 1838–1849. [CrossRef]

51. Lin, Y.W.; Bushman, J.D.; Yan, F.F.; Haidar, S.; MacMullen, C.; Ganguly, A.; Stanley, C.A.; Shyng, S.L. Destabilization of ATP-sensitive potassium channel activity by novel KCNJ11 mutations identified in congenital hyperinsulinism. *J. Biol. Chem.* **2008**, *283*, 9146–9156. [CrossRef]

52. Schulte, U.; Hahn, H.; Konrad, M.; Jeck, N.; Derst, C.; Wild, K.; Weidemann, S.; Ruppersberg, J.P.; Fakler, B.; Ludwig, J. pH gating of ROMK (K(ir)1.1) channels: Control by an Arg-Lys-Arg triad disrupted in antenatal Bartter syndrome. *Proc. Natl. Acad. Sci. USA* **1999**, *96*, 15298–15303. [CrossRef] [PubMed]

53. Scholl, U.I.; Choi, M.; Liu, T.; Ramaekers, V.T.; Häusler, M.G.; Grimmer, J.; Tobe, S.W.; Farhi, A.; Nelson-Williams, C.; Lifton, R.P. Seizures, sensorineural deafness, ataxia, mental retardation, and electrolyte imbalance (SeSAME syndrome) caused by mutations in KCNJ10. *Proc. Natl. Acad. Sci. USA.* **2009**, *106*, 5842–5847. [CrossRef] [PubMed]

54. Limberg, M.M.; Zumhagen, S.; Netter, M.F.; Coffey, A.J.; Grace, A.; Rogers, J.; Böckelmann, D.; Rinné, S.; Stallmeyer, B.; Decher, N.; et al. Non dominant-negative KCNJ2 gene mutations leading to Andersen-Tawil syndrome with an isolated cardiac phenotype. *Basic Res. Cardiol.* **2013**, *108*, 353. [CrossRef] [PubMed]

55. Fallen, K.; Banerjee, S.; Sheehan, J.; Addison, D.; Lewis, L.M.; Meiler, J.; Denton, J.S. The Kir channel immunoglobulin domain is essential for Kir1.1 (ROMK) thermodynamic stability, trafficking and gating. *Channels* **2009**, *3*, 57–68. [CrossRef]

56. O'Donnell, B.M.; Mackie, T.D.; Subramanya, A.R.; Brodsky, J.L. Endoplasmic reticulum-associated degradation of the renal potassium channel, ROMK, leads to type II Bartter syndrome. *J. Biol. Chem.* **2017**, *292*, 12813–12827. [CrossRef]

57. Károlyi, L.; Konrad, M.; Köckerling, A.; Ziegler, A.; Zimmermann, D.K.; Roth, B.; Wieg, C.; Grzeschik, K.H.; Koch, M.C.; Seyberth, H.W.; et al. Mutations in the gene encoding the inwardly-rectifying renal potassium channel, ROMK, cause the antenatal variant of Bartter syndrome: Evidence for genetic heterogeneity. International Collaborative Study Group for Bartter-like Syndromes. *Hum. Mol. Genet.* **1997**, *6*, 17–26.

58. Fodstad, H.; Swan, H.; Auberson, M.; Gautschi, I.; Loffing, J.; Schild, L.; Kontula, K. Loss-of-function mutations of the K$^+$ channel gene KCNJ2 constitute a rare cause of long QT syndrome. *J. Mol. Cell. Cardiol.* **2004**, *37*, 593–602. [CrossRef]

59. Davies, N.P.; Imbrici, P.; Fialho, D.; Herd, C.; Bilsland, L.G.; Weber, A.; Mueller, R.; Hilton-Jones, D.; Ealing, J.; Boothman, B.R.; et al. Andersen-Tawil syndrome: New potassium channel mutations and possible phenotypic variation. *Neurology* **2005**, *65*, 1083–1089. [CrossRef]

60. Lu, C.W.; Lin, J.H.; Rajawat, Y.S.; Jerng, H.; Rami, T.G.; Sanchez, X.; DeFreitas, G.; Carabello, B.; DeMayo, F.; Kearney, D.L.; et al. Functional and clinical characterization of a mutation in KCNJ2 associated with Andersen-Tawil syndrome. *J. Med. Genet.* **2006**, *43*, 653–659. [CrossRef]

61. Eckhardt, L.L.; Farley, A.L.; Rodriguez, E.; Ruwaldt, K.; Hammill, D.; Tester, D.J.; Ackerman, M.J.; Makielski, J.C. KCNJ2 mutations in arrhythmia patients referred for LQT testing: A mutation T305A with novel effect on rectification properties. *Heart Rhythm.* **2007**, *4*, 323–329. [CrossRef]

62. Tani, Y.; Miura, D.; Kurokawa, J.; Nakamura, K.; Ouchida, M.; Shimizu, K.; Ohe, T.; Furukawa, T. T75M-KCNJ2 mutation causing Andersen-Tawil syndrome enhances inward rectification by changing Mg^{2+} sensitivity. *J. Mol. Cell. Cardiol.* **2007**, *43*, 187–196. [CrossRef] [PubMed]

63. Snider, K.E.; Becker, S.; Boyajian, L.; Shyng, S.L.; MacMullen, C.; Hughes, N.; Ganapathy, K.; Bhatti, T.; Stanley, C.A.; Ganguly, A. Genotype and phenotype correlations in 417 children with congenital hyperinsulinism. *J. Clin. Endocrinol. Metab.* **2013**, *98*, 355–363. [CrossRef] [PubMed]

64. Mohnike, K.; Wieland, I.; Barthlen, W.; Vogelgesang, S.; Empting, S.; Mohnike, W.; Meissner, T.; Zenker, M. Clinical and genetic evaluation of patients with KATP channel mutations from the German registry for congenital hyperinsulinism. *Horm. Res. Paediatr.* **2014**, *81*, 156–168. [CrossRef] [PubMed]

65. Decher, N.; Renigunta, V.; Zuzarte, M.; Soom, M.; Heinemann, S.H.; Timothy, K.W.; Keating, M.T.; Daut, J.; Sanguinetti, M.C.; Splawski, I. Impaired interaction between the slide helix and the C-terminus of Kir2.1: A novel mechanism of Andersen syndrome. *Cardiovasc. Res.* **2007**, *75*, 748–757. [CrossRef] [PubMed]

66. Yoon, G.; Oberoi, S.; Tristani-Firouzi, M.; Etheridge, S.P.; Quitania, L.; Kramer, J.H.; Miller, B.L.; Fu, Y.H.; Ptácek, L.J. Andersen-Tawil syndrome: Prospective cohort analysis and expansion of the phenotype. *Am. J. Med. Genet. A* **2006**, *140*, 312–321. [CrossRef]

67. Ballester, L.Y.; Benson, D.W.; Wong, B.; Law, I.H.; Mathews, K.D.; Vanoye, C.G.; George, A.L. Jr. Trafficking-competent and trafficking-defective KCNJ2 mutations in Andersen syndrome. *Hum. Mutat.* **2006**, *27*, 388. [CrossRef]

68. Williams, D.M.; Lopes, C.M.; Rosenhouse-Dantsker, A.; Connelly, H.L.; Matavel, A.; O-Uchi, J.; McBeath, E.; Gray, D.A. Molecular basis of decreased Kir4.1 function in SeSAME/EAST syndrome. *J. Am. Soc. Nephrol.* **2010**, *21*, 2117–2129. [CrossRef]

69. Takeda, I.; Takahashi, T.; Ueno, H.; Morino, H.; Ochi, K.; Nakamura, T.; Hosomi, N.; Kawakami, H.; Hashimoto, K.; Matsumoto, M. Autosomal recessive Andersen–Tawil syndrome with a novel mutation L94P in Kir2.1. *Neurol. Clin. Neurosci.* **2013**, *1*, 131–137. [CrossRef]

70. Kuß, J.; Stallmeyer, B.; Goldstein, M.; Rinné, S.; Pees, C.; Zumhagen, S.; Seebohm, G.; Decher, N.; Pott, L.; Kienitz, M.C.; et al. Familial Sinus Node Disease Caused by a Gain of GIRK (G-Protein Activated Inwardly Rectifying K+ Channel) Channel Function. *Circ. Genom. Precis. Med.* **2019**, *12*, e002238. [CrossRef]

71. Derst, C.; Wischmeyer, E.; Preisig-Müller, R.; Spauschus, A.; Konrad, M.; Hensen, P.; Jeck, N.; Seyberth, H.W.; Daut, J.; Karschin, A. A hyperprostaglandin E syndrome mutation in *Kir1. 1* (renal outer medullary potassium) channels reveals a crucial residue for channel function in Kir1. 3 channels. *J. Biol. Chem.* **1998**, *273*, 23884–23891. [CrossRef]

72. Cheng, C.J.; Sung, C.C.; Wu, S.T.; Lin, Y.C.; Sytwu, H.K.; Huang, C.L.; Lin, S.H. Novel KCNJ5 mutations in sporadic aldosterone-producing adenoma reduce Kir3.4 membrane abundance. *J. Clin. Endocrinol. Metab.* **2015**, *100*, E155–E163. [CrossRef] [PubMed]

73. Gélinas, R.; El Khoury, N.; Chaix, M.A.; Beauchamp, C.; Alikashani, A.; Ethier, N.; Boucher, G.; Villeneuve, L.; Robb, L.; Latour, F.; et al. Characterization of a Human Induced Pluripotent Stem Cell-Derived Cardiomyocyte Model for the Study of Variant Pathogenicity: Validation of a KCNJ2 Mutation. *Circ. Cardiovasc. Genet.* **2017**, *10*, e001755. [CrossRef] [PubMed]

74. Cheng, C.J.; Lin, S.H.; Lo, Y.F.; Yang, S.S.; Hsu, Y.J.; Cannon, S.C.; Huang, C.L. Identification and functional characterization of Kir2.6 mutations associated with non-familial hypokalemic periodic paralysis. *J. Biol. Chem.* **2011**, *286*, 27425–27435. [CrossRef] [PubMed]

75. Lee, S.J.; Ren, F.; Zangerl-Plessl, E.M.; Heyman, S.; Stary-Weinzinger, A.; Yuan, P.; Nichols, C.G. Structural basis of control of inward rectifier Kir2 channel gating by bulk anionic phospholipids. *J. Gen. Physiol.* **2016**, *148*, 227–237. [CrossRef]

76. Tanemoto, M.; Abe, T.; Uchida, S.; Kawahara, K. Mislocalization of K$^+$ channels causes the renal salt wasting in EAST/SeSAME syndrome. *FEBS Lett.* **2014**, *588*, 899–905. [CrossRef]

77. Pattnaik, B.R.; Tokarz, S.; Asuma, M.P.; Schroeder, T.; Sharma, A.; Mitchell, J.C.; Edwards, A.O.; Pillers, D.A. Snowflake vitreoretinal degeneration (SVD) mutation R162W provides new insights into Kir7.1 ion channel structure and function. *PLoS ONE* **2013**, *8*, 71744. [CrossRef]

78. Sergouniotis, P.I.; Davidson, A.E.; Mackay, D.S.; Li, Z.; Yang, X.; Plagnol, V.; Moore, A.T.; Webster, A.R. Recessive mutations in KCNJ13, encoding an inwardly rectifying potassium channel subunit, cause leber congenital amaurosis. *Am. J. Hum. Genet.* **2011**, *89*, 183–190. [CrossRef]

79. McDonald, S.K.; Levitz, T.S.; Valiyaveetil, F.I. A Shared Mechanism for the Folding of Voltage-Gated K+ Channels. *Biochemistry* **2019**, *58*, 1660–1671. [CrossRef]

80. Cho, H.C.; Tsushima, R.G.; Nguyen, T.T.; Guy, H.R.; Backx, P.H. Two critical cysteine residues implicated in disulfide bond formation and proper folding of Kir2. 1. *Biochemistry* **2000**, *39*, 4649–4657. [CrossRef]

81. Hristova, K.; Wimley, W.C. A look at arginine in membranes. *J. Membr. Biol.* **2011**, *239*, 49–56. [CrossRef]

 *biomolecules*

Review

Mechanisms and Alterations of Cardiac Ion Channels Leading to Disease: Role of Ankyrin-B in Cardiac Function

Holly C. Sucharski [1,2,†], Emma K. Dudley [1,2,†], Caullin B.R. Keith [1,2], Mona El Refaey [1,2], Sara N. Koenig [1,2,*] and Peter J. Mohler [1,2]

[1]   Dorothy M. Davis Heart and Lung Research Institute, The Ohio State University Wexner Medical Center, Columbus, OH 43210, USA; holly.sucharski@osumc.edu (H.C.S.); emma.dudley@osumc.edu (E.K.D.); Caullin.Keith@osumc.edu (C.B.R.K.); Mona.elrefaey@osumc.edu (M.E.R.); peter.mohler@osumc.edu (P.J.M.)

[2]   Departments of Physiology and Cell Biology and Internal Medicine, Division of Cardiovascular Medicine, The Ohio State University College of Medicine and Wexner Medical Center, Columbus, OH 43210, USA

*   Correspondence: sara.koenig@osumc.edu; Tel.: +1-614-366-0510

†   Denotes equal contribution.

Received: 2 January 2020; Accepted: 28 January 2020; Published: 31 January 2020

Abstract: Ankyrin-B (encoded by *ANK2*), originally identified as a key cytoskeletal-associated protein in the brain, is highly expressed in the heart and plays critical roles in cardiac physiology and cell biology. In the heart, ankyrin-B plays key roles in the targeting and localization of key ion channels and transporters, structural proteins, and signaling molecules. The role of ankyrin-B in normal cardiac function is illustrated in animal models lacking ankyrin-B expression, which display significant electrical and structural phenotypes and life-threatening arrhythmias. Further, ankyrin-B dysfunction has been associated with cardiac phenotypes in humans (now referred to as "ankyrin-B syndrome") including sinus node dysfunction, heart rate variability, atrial fibrillation, conduction block, arrhythmogenic cardiomyopathy, structural remodeling, and sudden cardiac death. Here, we review the diverse roles of ankyrin-B in the vertebrate heart with a significant focus on ankyrin-B-linked cell- and molecular-pathways and disease.

Keywords: ankyrin-B; *ANK2*; ion channels; cardiovascular disease

---

1. Introduction: Ankyrin Proteins

The ankyrin family of polypeptides was first identified in the erythrocyte plasma membrane in 1979 by Bennett and Stenbuck [1]. Following this discovery, ankyrin was discovered in various organs and cell types, including brain [2–4] and myogenic cells [5,6]. We now know that ankyrins are derived from three ankyrin genes: *ANK1, ANK2,* and *ANK3* [7–9]. *ANK1* encodes ankyrin-R (AnkR), *ANK2* encodes ankyrin-B (AnkB), and *ANK3* encodes ankyrin-G (AnkG) (Table 1).

Table 1. Brief summary of ankyrin family proteins: ankyrin-R, ankyrin-B, ankyrin-G.

	Ankyrin-R	Ankyrin-B	Ankyrin-G
Tissue Expression	erythrocytes [1], myelinated axons [10], striated muscle [11]	ubiquitously expressed, cardiomyocytes (T-tubules, SR, plasma membrane) [12], neurons [8]	ubiquitously expressed, neurons (AIS, and nodes of Ranvier) [13], cardiomyocytes (intercalated disc) [14]
Examples of Binding Partners	CD44 [15], NKA [16], Rh type A glycoprotein [17], obscurin [11]	PP2A [12,13], NCX [12], NKA [18], Kir6.2 [12,13], $Ca_V1.3$ [19], βII-spectrin [20]	$Na_V1.6$, βIV-spectrin, L1CAMs [1,21,22], plakophilin-2 [23] $Na_V1.5$ [14]
Isoforms	sAnk1.5, 1.6, 1.7, and 1.9 [11]	AnkB-188 and AnkB-212 [24]. Giant AnkB (440-kD)	Giant AnkG (480-kD) [25]
Disease associated with variants	hereditary spherocytosis [26]	Ankyrin B syndrome: SCD, SND, AF, LQTS, VT, bradycardia, syncope [12], ARVC [27]	Brugada syndrome [12], dilated cardiomyopathy [28], cognitive disabilities [29]

SR = sarcoplasmic reticulum, AIS = axon initial segments, NKA = Na^+/K^+ ATPase, PP2A = protein phosphatase 2A, NCX = Na^+/Ca^{2+} exchanger, Kir6.2 = inward rectifier potassium channel, $Ca_V1.3$ = voltage-gated calcium channel, L1CAMs = L1 family of neural cell adhesion molecules, SCD = sudden cardiac death, SND = sinus node disease, AF = atrial fibrillation, LQTS = long QT syndrome, VT = ventricular tachycardia, ARVC = arrhythmogenic right ventricular cardiomyopathy.

Ankyrin-R is primarily expressed in erythrocytes and was the first ankyrin identified as an adaptor protein [1], linking erythrocyte membrane proteins to the actin-based cytoskeleton [30]. AnkR is important for the structural integrity and organization of the erythrocyte membrane. Moreover, loss-of-function variants in human *ANK1* have been linked to ~50% of hereditary spherocytosis cases [31], a complex form of hemolytic anemia that affects 1:2000 people of Northern European descent [26] and results from a loss of membrane surface tension in red blood cells [32]. AnkR regulates erythrocyte membrane expression of CD44 [15], Na^+/K^+ ATPase (NKA) [16], and the Rh type A glycoprotein [17]. Although AnkR is primarily expressed in erythrocytes, AnkR is also expressed in myelinated axons [10]. *ANK1* also encodes four small ankyrin-1 isoforms (sAnk1.5, 1.6, 1.7, and 1.9) that are highly expressed in striated muscle. These isoforms bind obscurin and stabilize the sarcoplasmic reticulum in striated muscle [11].

Ankyrin-B was first identified in the brain [8] and has since been identified as a critical adaptor and scaffolding protein in the heart, that mediates the interaction of integral membrane proteins with the spectrin-actin cytoskeletal network [13,33]. *ANK2* encodes multiple isoforms that may contribute to disease. AnkB will be reviewed in detail below.

Ankyrin-G plays an important role across multiple excitable tissues. In the brain, AnkG links integral membrane proteins with the actin/spectrin-based membrane skeleton at axon initial segments (AIS) including $Na_V1.6$, βIV spectrin, and L1CAMs [21,22,33]. In the heart, AnkG is required for localization of $Na_V1.5$ and CaMKII to the cardiomyocyte intercalated disc [13,14,34]. In mice selectively lacking AnkG expression in cardiomyocytes, βIV-spectrin and $Na_V1.5$ expression and localization are disrupted, and voltage-gated Na_V channel activity (I_{Na}) is significantly decreased. These animals experience a reduction in heart rate, impaired atrioventricular conduction, increased PR intervals, and increased QRS intervals [14]. Further, AnkG cKO mice display arrhythmias in response to adrenergic stimulation. In humans, an *SCN5A* variant in the AnkG-binding motif of $Na_V1.5$ has been associated with Brugada syndrome and arrhythmia [12]. This same variant is a loss-of-function variant when expressed in primary cardiomyocytes. Similar to other ankyrin genes, *ANK3* encodes multiple isoforms of AnkG. Giant AnkG is a 480-kD protein required for proper AIS and node of Ranvier assembly due to the clustering of Na_V channels [35]. Human variants affecting 480-kD AnkG are associated with severe cognitive disability [29]. The role of Giant AnkG isoforms in the heart is currently unknown and is an important area for future research.

AnkB and AnkG are ubiquitously expressed, but their functions are distinct. Although AnkG plays a crucial role in the brain, variants in AnkG have been connected to Brugada syndrome [12] and, more recently, dilated cardiomyopathy [28]. Although AnkB and AnkG have similar structures, AnkG partners with proteins at the intercalated disc, including plakophilin-2 [23] and $Na_V1.5$ [14], while AnkB is crucial for the expression and localization of ion channels at the sarcoplasmic reticulum, transverse-tubules, and plasma membrane [12]. However, Roberts et al. recently identified small populations of AnkB at the intercalated disc [27]. Cardiomyocytes from mice heterozygous for a null mutation in ankyrin-B display mislocalization and a decrease in expression of Na^+/Ca^{2+} exchanger (NCX) and Na^+/K^+-ATPase (NKA) [36]. Further, Roberts et al. demonstrated that β-catenin is a novel AnkB-binding partner, where β-catenin localization is disrupted in individuals with *ANK2* variants who presented with arrhythmogenic right ventricular cardiomyopathy (ARVC) [27]. Importantly, ankyrins -G and -B retain non-overlapping, non-compensatory functions despite their similarity in sequence. Distinct from AnkG-associated disease, variants in AnkB are tied to a specific set of clinical phenotypes, including susceptibilities to sinus node dysfunction and acquired heart diseases such as atrial fibrillation [12] and heart failure [37]. Ankyrin specificity, at least in part, is attributed to an autoinhibitory linker peptide between the membrane-binding domain (MBD) and spectrin-binding domain (SBD), which prevents AnkB from binding with protein partners [38]. Further specificity is attributed to key roles of the divergent C-terminal domains of AnkB and AnkG. Additional mechanisms underlying ankyrin specificity in vivo are a key area for future research.

2. Ankyrin-B Isoforms

Similar to *ANK1* and *ANK3*, *ANK2* generates multiple gene products. The diversity of known and potential gene products is large and includes splice products defined as small, canonical, and Giant AnkB isoforms [13]. In fact, The National Center for Biotechnology Information gene database (NCBI Gene) lists 49 transcript variants that match the known RefSeq (NM) and 20 transcript variants that match the model RefSeq (XM), yet most of these transcript variants have not been identified in tissue or cells.

AnkB-188 and AnkB-212 are two ankyrin-B isoforms present in the heart [24]. AnkB-188 is expressed in human ventricular cardiomyocytes and regulates NCX expression, whereas AnkB-212 is expressed in cardiomyocytes and skeletal muscle, is localized to the M-line, and exclusively interacts with obscurin [24]. Furthermore, 440kD AnkB (Giant AnkB) is the result of the insertion of a 6.4 kb exon between the SBD and death domain (DD). To date, expression of Giant AnkB has only been studied in neurons and was recently associated with autism [25]. Giant AnkB regulates the expression and localization of L1CAMs in neurons and predominates in unmyelinated axons to control axonal branching. A mouse model deficient in Giant AnkB shows increased axonal branching and a transient increase in excitatory synapses during postnatal development [39]. Although there are a host of ankyrin-B isoforms in the heart, the primary isoform is canonical 220-kD AnkB.

Canonical AnkB is an adaptor protein that acts as a pivotal regulator in the localization and organization of ion channels, structural proteins, signaling molecules, and adaptor proteins [40,41]. Deficiency in AnkB results in mislocalization and altered expression of multiple membrane and cytoskeletal proteins, including NCX [12], NKA [18], ATP-sensitive inward rectifier K^+ channel (Kir6.2) [13], voltage-gated calcium channel ($Ca_V1.3$) [19], and βII-spectrin [20], causing ion imbalance and dysregulation of cellular signaling [18].

Animal models have been fundamental in illustrating the role of AnkB in the heart. $AnkB^{+/-}$ myocytes display reduced expression and improper localization of AnkB-binding partners, as well as strong electrical phenotypes including extrasystoles, early afterdepolarizations (EADs), and delayed afterdepolarizations (DADs) following adrenergic stimulation. The mechanism underlying AnkB-associated arrhythmias is attributed to improper calcium handling via disrupted organization of AnkB-binding partners [36]. Through the use of a canine cardiomyocyte model, Chu et al. predicted that mislocalization of the NKA and its subsequent uncoupling from the NCX in $AnkB^{+/-}$ cardiomyocytes disturbs Ca^{2+} and Na^+ currents to predispose the cell to action potential prolongation [42].

3. Ankyrin-B Structure and Binding Partners

Canonical ankyrin proteins share a similar structure composed of an MBD, SBD, and a regulatory domain (RD) that is comprised of a death domain (DD) and C-terminal domain (CTD) (Figure 1). The MBD of AnkB is divided into four subdomains composed of 24 *ANK* repeats, which are defined by their repeated alpha-helical structure [43]. *ANK* repeats are not specific to the three canonical ankyrin proteins and are present across a range of functionally diverse proteins including SHANK, BARD1, and ANKRD. The AnkB MBD regulates the localization of ion channels and transporters (Figure 2 and Table 2).

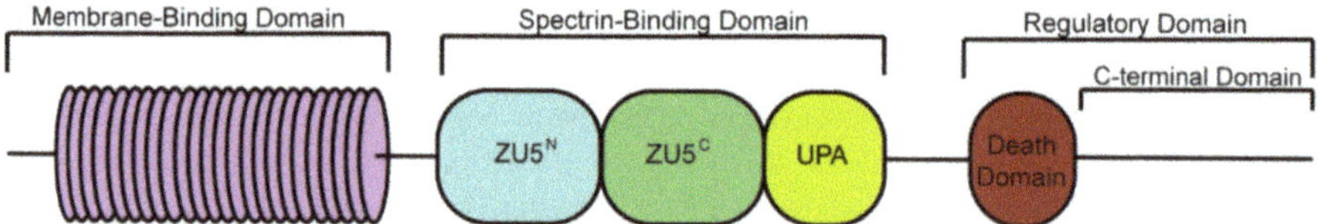

Figure 1. Structure of canonical ankyrin-B. Canonical ankyrin proteins share four domains: a membrane-binding domain (MBD), spectrin-binding domain (SBD), death domain (DD), and C-terminal domain (CTD). The MBD consists of 24 *ANK* repeats that are defined by their secondary structure and aid in protein folding regulation. The SBD consists of ZU5N, ZU5C, and UPA domains that are important for binding βII-spectrin and supporting cardiomyocyte structure. The DD and CTD comprise the regulatory domain.

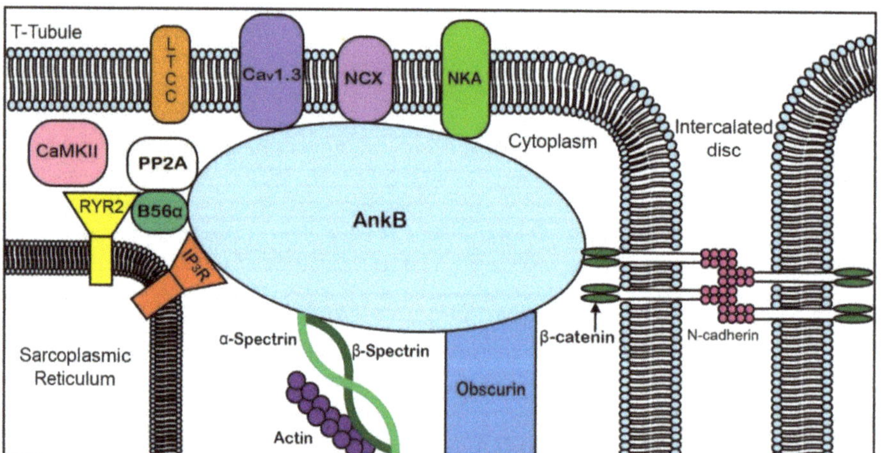

Figure 2. Representative diagram of ankyrin-B-binding partners to emphasize the importance of AnkB in the localization of ion channels, transporters, pumps, and structural proteins for proper cardiomyocyte function.

Table 2. Ankyrin-B-binding partners in the heart.

Membrane-Binding Domain		Spectrin-Binding Domain	Regulatory Domain
Ion channels	**Transporters/Pumps**	β-spectrin	HSP40
IP3R	Anion Exchanger	PP2A	Obscurin
Ca$_V$1.3	Na/Ca Exchanger		Ankyrin MBD
Kir6.2	Na/K ATPase		
Structural	**Cell adhesion**		
Tubulin β-catenin	L1CAMs		
	β-dystroglycan		
	Dystrophin		

IP3R = 1,4,5 inositol trisphosphate receptor, Ca$_V$1.3 = voltage-gated Ca^{2+} channel, Kir6.2 = ATP-sensitive inward rectifier K$^+$ channel, PP2A = protein phosphatase 2A, HSP40 = heat shock protein 40, MBD = membrane binding domain, L1CAMs = L1 family of neural cell adhesion molecules.

The SBD is a highly conserved region of most ankyrin proteins, where the AnkB SBD is composed of a ZU5^N-ZU5^C-UPA tandem structure. The ZU5^N domain binds directly with the spectrin cytoskeleton to aid in the structural integrity of the myocyte [44]. Critical to calcium handling in the heart, the ankyrin ZU5^C region of the AnkB-SBD associates with the B56α subunit of protein phosphatase 2A (PP2A), an interaction that localizes PP2A with a primary target, RyR2 [12,13]. The *ANK2* p.Q1283H variant within ZU5c was recently associated with arrhythmia, potentially due to loss of PP2A activity and altered phosphorylation of RyR2 [45].

The AnkB DD, composed of six helices that resemble the death domain of apoptotic-related proteins, is proposed to play a role in the auto-inhibitory functions of AnkB through the binding of the DD to the UPA region of SBD [44]. Additionally, a yeast two-hybrid screen and subsequent co-immunoprecipitation revealed that the AnkB DD binds RAB GTPase activating protein-1 like (RABGAP1L), which is involved in intracellular membrane trafficking within many different cell types, including heart tissue [46]. Although the regulatory domain is the least conserved domain of AnkB, the majority of disease-associated variants have been identified in this region. The DD variant, V1516D, was identified in four individuals with various heart conditions such as atrial fibrillation, drug-induced long QT syndrome (LQTS), exercise-induced ventricular tachycardia, bradycardia, syncope, or a combination of phenotypes [12].

Finally, the AnkB CTD is characterized by its elongated structure and its high composition of charged amino acids [12]. It is highly divergent and interacts with obscurin [47]. The CTD structure allows for inter-domain interactions that are crucial to isoform-specific functions. Intramolecular interactions between the AnkB CTD and MBD are proposed to regulate protein binding interactions. Ankyrin auto-regulatory activity was originally identified in AnkR and its splice variant AnkR 2.2 [48,49]. Within AnkB, yeast two-hybrid studies identified essential amino acid sequences of the CTD and MBD that are required for binding; Arg 37 and Arg 40 of the *ANK* repeat 1 within the MBD and the three amino acid sequence EED within the CTD are shown to facilitate inter-domain binding [12]. Abolishment of this interaction inhibits AnkB-specific targeting of inositol trisphosphate receptor (IP3R) to the sarcoplasmic reticulum, therefore demonstrating that inter-domain interactions are responsible for ankyrin function through regulation of proper protein interactions. Interestingly, however, these amino acid sequences are not required for binding of AnkB MBD with the IP3R, raising questions about the role of the CTD in localization and expression of AnkB-binding partners within the heart.

4. Ankyrin-B Variants in Cardiovascular Disease

AnkB variants have been identified in all four domains of the AnkB protein and are linked to a spectrum of cardiovascular phenotypes. AnkB is classically associated with human arrhythmia syndromes, many of which demonstrate incomplete penetrance and variable expressivity [12,50–52]. In fact, it is likely that secondary genetic, lifestyle, and/or environmental factors are necessary to cause disease. Ankyrin-B syndrome, originally classified as long QT syndrome type 4, is a heritable arrhythmogenic disease that is the result of loss-of-function mutations in *ANK2*. A p.E1425G variant was discovered in a French family suffering from sinus bradycardia, atrial fibrillation, and sudden cardiac death and was the first to be implicated in AnkB syndrome [12,53].

Although the p.E1425G variant is localized to the AnkB regulatory domain, loss-of-function variants in all four ankyrin domains are now associated with AnkB syndrome [12,54]. Notably, these variants show a range of clinical severity and phenotypes, including torsades de pointes (TdP), ventricular tachycardia, and long QT syndrome, a variability that is reflected at the cellular level [12]. This inconsistency of phenotype puts into the question the mechanics behind AnkB-associated diseases that have yet to be fully elucidated. Although the AnkB MBD is imperative for AnkB interactions and function within the heart, the first disease-associated loss-of-function variant in this domain, p.S646F, was only recently identified within the First Nations of Northern British Columbia [54]. Individuals with this variant also exhibited congenital heart defects, Wolff–Parkinson–White syndrome, and cardiomyopathy. These compounded symptoms are the result of improper NCX localization, subsequent Ca^{2+} overload, and possible disruption of pacemaking activity [12,36].

Variants within the SBD pose distinct mechanisms of disease. Several AnkB variants including the p.R990Q [20], p.A1000P, and p.DAR976AAA are present within the highly conserved ZU5^N region of the SBD that directly binds βII-spectrin, disrupting this interaction. Importantly, AnkB co-immunoprecipitates with a larger complex composed of βII-spectrin, NKA, and NCX [20]. Although p.A1000P and p.DAR976AAA demonstrated normal AnkB activity with altered βII-spectrin binding, the p.R990Q variant showed altered AnkB functionality, including an inability to rescue NCX localization in AnkB knockout myocytes and severe arrhythmia attributed to disruption of the AnkB-spectrin interaction [20]. Notably, the p.Q1283H variant within the ZU5^C region disrupts PP2A activity via loss of B56α targeting, which increases RyR2 phosphorylation and disrupts the calcium dynamics associated with excitation-contraction (EC) coupling [45]. Notably, alterations in RyR2 activity are associated with dilated cardiomyopathy and offer an additional area for future investigation.

Variants in *ANK2* have also been associated with sinus node disease (SND) [12] and most recently with arrhythmogenic cardiomyopathy (ACM) [27]. In fact, the p.E1425G variant segregated with sinus node disease with nearly complete penetrance [45]. Similar to *ANK2* loss of function mutations, *ANK2* transection in chromosome 4, leading to *ANK2* haploinsufficiency, was associated with ankyrin-B syndrome [55].

Although AnkB is commonly associated with arrhythmias, variants in *ANK2* have also been associated with structural heart disease. Lopes et al. found that individuals with *ANK2* variants had a greater maximum wall thickness in the left ventricle, in hypertrophic cardiomyopathy [50]. AnkB has also been linked to acquired heart disease and tissue remodeling following infarct in canine animal models [12]. Following coronary artery occlusion, AnkB mRNA and protein levels decrease in the cardiomyocytes found at the infarct border zone (BZ), as do common AnkB-binding partners including NCX and NKA. These findings provide new insight into the role of AnkB in heart failure and possibly for the cardiac remodeling that supports the creation of arrhythmogenic substrates at the border zone [37].

Recently, *ANK2* variants have been identified in individuals with arrhythmogenic right ventricular cardiomyopathy (ARVC) [27], a disease characterized by a severe structural and electrical cardiac phenotype that involves the fibrofatty replacement of healthy myocardium, malignant arrhythmias, and even sudden cardiac death [56]. The AnkB-p.Glu1458Gly variant was linked to AnkB syndrome but was also identified in a family with AnkB syndrome found to have ARVC at autopsy [27]. A larger screen that encompassed AnkB variants in ACM revealed a loss-of-function variant, AnkB-p.Met1988Thr, which segregated with ARVC-affected family members, where staining of the ventricle tissue showed reduced levels of NCX at the plasma membrane and abnormal Z-line targeting [27]. In order to model the ACM phenotype seen in these patients with loss-of-function AnkB variants and ACM, a cardiomyocyte-specific AnkB knockout mouse was generated (as AnkB null mice die shortly after birth), which developed a phenotype similar to that of human ACM, including dramatic structural abnormalities, biventricular dilation, reduced ejection fraction, cardiac fibrosis, premature death, and exercise-induced death [27]. Although the desmosome was preserved in these mice, β-catenin localization was altered, and β-catenin was demonstrated as a binding partner for the AnkB MBD. Interestingly, when these mice were treated with GSK-3β-inhibitor, a pharmacological activator of β-catenin, it prevented and partially reversed the ARVC phenotype found in these mice [27], providing a hopeful outlook on developing new therapies for patients with AnkB variant-driven ARVC.

5. Future Implications

ANK2 variants are associated with cardiovascular phenotypes including sinus node disease, atrial fibrillation, heart rate variability, catecholaminergic polymorphic ventricular tachycardia (CPVT), ACM, cardiomyopathy, syncope, and sudden cardiac death. We strongly predict that disease penetrance and severity will ultimately be predicated by the interaction between genetic (known as well as currently unknown variants, deletions, etc.) and environmental factors. A key future area of research is to understand the impact of secondary variants and acquired/environmental factors (e.g., ischemia, catecholamines) on ankyrin-B stability and disease.

Despite their variability, the majority of treatment options for AnkB-associated conditions continue to mitigate the clinical symptoms of disease as opposed to their molecular causes. Currently, SND, ARVC, and AnkB syndrome necessitate symptom-dependent approaches ranging from β-blocker administration to pacemaker implantation [12,37,56]. Only recently was SB-216763, a GSK-3β inhibitor, identified to both prevent and reverse ARVC in mice by targeting a specific interaction between AnkB and β-catenin through the Wnt signaling pathway [27], but this therapeutic strategy has not yet been tested. Although the authors posit that the drug may function by altering β-catenin phosphorylation levels, the exact mechanics of its action remain unclear, which highlights the need for future examination. Furthermore, these findings address the structural phenotype of ARVC that is largely absent in SND and AnkB syndrome, again establishing the need for more disease-specific therapeutic options.

The clinical manifestations of AnkB-attributed pathologies are also intriguingly patient specific. For example, cases of ARVC and AnkB syndrome, which comprise disparate structural and electrical phenotypes, have both been linked to the AnkB E1425G variant [27]. This genetic overlap demonstrates that individual AnkB variants are not likely the sole determinants of a given clinical phenotype, but rather are compounded by environmental factors that work in conjunction with additional genetic factors to produce disease variability. Further study is necessary to elucidate the synergistic interactions and intricate molecular pathways that could act as novel therapeutic targets in AnkB-associated illness.

In summary, AnkB is critical for the expression, targeting, and regulation of multiple proteins involved in cardiac excitability, structure, and signaling. AnkB is a pivotal regulator of ion channels and transporters (e.g., NCX and NKA), structural proteins (e.g., βII-spectrin and β-catenin), and calcium regulatory proteins (e.g., RyR2 and PP2A) (Figure 2). Together, the diverse roles of AnkB contribute to EC coupling, cytoskeletal integrity, and signaling pathways required for proper cardiomyocyte function. Variants in AnkB are associated with arrhythmogenic diseases that include both structural and electrical dysfunction. Although AnkB isoforms have been shown to have unique functions in the heart (AnkB-188 and AnkB-212), it is not known if variants in these isoforms contribute to disease. Since AnkB-188 plays a role in localization of NCX and AnkB-212 binds obscurin, variants may lead to M-line disruption in cardiomyocyte conduction and contraction, as knockdown of either isoform resulted in arrhythmic contraction in vitro [24]. Although Giant AnkB has only been studied in the brain, where variants may increase susceptibility to cognitive disorders and autism, its expression and role in heart disease has yet to be determined. Understanding the role of AnkB variants in disease, along with other variants that contribute to the phenotypes described here, is crucial to developing new treatment strategies for complex human disease.

Author Contributions: All authors listed have made substantial direct and intellectual contribution to the work, and approved it for publication.

Funding: This work was supported by NIH grants HL135754 and HL134824 to PJM, HL146969 to MER, and by a grant from the Ohio State Frick Center for Heart Failure and Arrhythmia, the Linda and Joe Chlapaty Center for Atrial Fibrillation, and JB Project.

Conflicts of Interest: The authors declare no conflict of interest.

References

1. Bennett, V.; Stenbuck, P.J. Identification and partial purification of ankyrin, the high affinity membrane attachment site for human erythrocyte spectrin. *J. Biol. Chem.* **1979**, *254*, 2533–2541. [PubMed]
2. Bennett, V.; Davis, J.; Fowler, W.E. Immunoreactive forms of erythrocyte spectrin and ankyrin in brain. *Philos. Trans. R. Soc. Lond. Ser. Biol. Sci.* **1982**, *299*, 301–312. [CrossRef] [PubMed]
3. Bennett, V.; Davis, J. Spectrin and ankyrin in brain. *Cell Motil.* **1983**, *3*, 623–633. [CrossRef] [PubMed]
4. Davis, J.Q.; Bennett, V. Brain ankyrin. Purification of a 72,000 Mr spectrin-binding domain. *J. Biol. Chem.* **1984**, *259*, 1874–1881.

5. Moon, R.T.; Ngai, J.; Wold, B.J.; Lazarides, E. Tissue-specific expression of distinct spectrin and ankyrin transcripts in erythroid and nonerythroid cells. *J. Cell Biol.* **1985**, *100*, 152–160. [CrossRef]

6. Nelson, W.J.; Lazarides, E. Posttranslational control of membrane-skeleton (ankyrin and alpha beta-spectrin) assembly in early myogenesis. *J. Cell Biol.* **1985**, *100*, 1726–1735. [CrossRef]

7. Otto, E.; Kunimoto, M.; McLaughlin, T.; Bennett, V. Isolation and characterization of cDNAs encoding human brain ankyrins reveal a family of alternatively spliced genes. *J. Cell Biol.* **1991**, *114*, 241–253. [CrossRef]

8. Davis, J.Q.; Bennett, V. Brain ankyrin. A membrane-associated protein with binding sites for spectrin, tubulin, and the cytoplasmic domain of the erythrocyte anion channel. *J. Biol. Chem.* **1984**, *259*, 13550–13559.

9. Lambert, S.; Yu, H.; Prchal, J.T.; Lawler, J.; Ruff, P.; Speicher, D.; Cheung, M.C.; Kan, Y.W.; Palek, J. cDNA sequence for human erythrocyte ankyrin. *Proc. Natl. Acad. Sci. USA* **1990**, *87*, 1730–1734. [CrossRef]

10. Ho, T.S.-Y.; Zollinger, D.R.; Chang, K.-J.; Xu, M.; Cooper, E.C.; Stankewich, M.C.; Bennett, V.; Rasband, M.N. A hierarchy of ankyrin-spectrin complexes clusters sodium channels at nodes of Ranvier. *Nat. Neurosci.* **2014**, *17*, 1664–1672. [CrossRef] [PubMed]

11. Pierantozzi, E.; Szentesi, P.; Al-Gaadi, D.; Oláh, T.; Dienes, B.; Sztretye, M.; Rossi, D.; Sorrentino, V.; Csernoch, L. Calcium Homeostasis Is Modified in Skeletal Muscle Fibers of Small Ankyrin1 Knockout Mice. *Int. J. Mol. Sci.* **2019**, *20*, 3361. [CrossRef] [PubMed]

12. Curran, J.; Mohler, P.J. Coordinating electrical activity of the heart: Ankyrin polypeptides in human cardiac disease. *Expert Opin. Targets* **2011**, *15*, 789–801. [CrossRef]

13. Cunha, S.R.; Mohler, P.J. Ankyrin-based cellular pathways for cardiac ion channel and transporter targeting and regulation. *Semin. Cell Dev. Biol.* **2011**, *22*, 166–170. [CrossRef] [PubMed]

14. Makara, M.A.; Curran, J.; Little, S.C.; Musa, H.; Polina, I.; Smith, S.A.; Wright, P.J.; Unudurthi, S.D.; Snyder, J.; Bennett, V.; et al. Ankyrin-G coordinates intercalated disc signaling platform to regulate cardiac excitability in vivo. *Circ. Res.* **2014**, *115*, 929–938. [CrossRef]

15. Kalomiris, E.L.; Bourguignon, L.Y. Mouse T lymphoma cells contain a transmembrane glycoprotein (GP85) that binds ankyrin. *J. Cell Biol.* **1988**, *106*, 319–327. [CrossRef] [PubMed]

16. Nelson, W.J.; Veshnock, P.J. Ankyrin binding to (Na$^+$ + K$^+$)ATPase and implications for the organization of membrane domains in polarized cells. *Nature* **1987**, *328*, 533–536. [CrossRef]

17. Nicolas, V.; Le Van Kim, C.; Gane, P.; Birkenmeier, C.; Cartron, J.-P.; Colin, Y.; Mouro-Chanteloup, I. Rh-RhAG/Ankyrin-R, a New Interaction Site between the Membrane Bilayer and the Red Cell Skeleton, Is Impaired by Rhnull-associated Mutation. *J. Biol. Chem.* **2003**, *278*, 25526–25533. [CrossRef]

18. Skogestad, J.; Aronsen, J.M.; Tovsrud, N.; Wanichawan, P.; Hougen, K.; Stokke, M.K.; Carlson, C.R.; Sjaastad, I.; Sejersted, O.M.; Swift, F. Coupling of the Na+/K+-ATPase to Ankyrin B controls Na+/Ca2+ exchanger activity in cardiomyocytes. *Cardiovasc. Res.* **2019**. [CrossRef]

19. Wolf, R.M.; Glynn, P.; Hashemi, S.; Zarei, K.; Mitchell, C.C.; Anderson, M.E.; Mohler, P.J.; Hund, T.J. Atrial fibrillation and sinus node dysfunction in human ankyrin-B syndrome: A computational analysis. *Am. J. Physiol. Heart Circ. Physiol.* **2013**, *304*, H1253–H1266. [CrossRef]

20. Smith, S.A.; Sturm, A.C.; Curran, J.; Kline, C.F.; Little, S.C.; Bonilla, I.M.; Long, V.P.; Makara, M.; Polina, I.; Hughes, L.D.; et al. Dysfunction in the βII spectrin-dependent cytoskeleton underlies human arrhythmia. *Circulation* **2015**, *131*, 695–708. [CrossRef]

21. Kordeli, E.; Lambert, S.; Bennett, V. Ankyrin: A new ankyrin gene with neural-specific isoforms localized at the axonal initial segment and node of ranvier. *J. Biol. Chem.* **1995**, *270*, 2352–2359. [CrossRef] [PubMed]

22. Jenkins, S.M.; Bennett, V. Ankyrin-G coordinates assembly of the spectrin-based membrane skeleton, voltage-gated sodium channels, and L1 CAMs at Purkinje neuron initial segments. *J. Cell Biol.* **2001**, *155*, 739–746. [CrossRef] [PubMed]

23. Sato, P.Y.; Coombs, W.; Lin, X.; Nekrasova, O.; Green, K.J.; Isom, L.L.; Taffet, S.M.; Delmar, M. Interactions between ankyrin-G, Plakophilin-2, and Connexin43 at the cardiac intercalated disc. *Circ. Res.* **2011**, *109*, 193–201. [CrossRef] [PubMed]

24. Wu, H.C.; Yamankurt, G.; Luo, J.; Subramaniam, J.; Hashmi, S.S.; Hu, H.; Cunha, S.R. Identification and characterization of two ankyrin-B isoforms in mammalian heart. *Cardiovasc. Res.* **2015**, *107*, 466–477. [CrossRef] [PubMed]

25. Kunimoto, M.; Otto, E.; Bennett, V. A new 440-kD isoform is the major ankyrin in neonatal rat brain. *J. Cell Biol.* **1991**, *115*, 1319–1331. [CrossRef]

26. Perrotta, S.; Gallagher, P.G.; Mohandas, N. Hereditary spherocytosis. *Lancet (Lond. Engl.)* **2008**, *372*, 1411–1426. [CrossRef]

27. Roberts, J.D.; Murphy, N.P.; Hamilton, R.M.; Lubbers, E.R.; James, C.A.; Kline, C.F.; Gollob, M.H.; Krahn, A.D.; Sturm, A.C.; Musa, H.; et al. Ankyrin-B dysfunction predisposes to arrhythmogenic cardiomyopathy and is amenable to therapy. *J. Clin. Investig.* **2019**, *129*, 3171–3184. [CrossRef]

28. Makara, M.A.; Curran, J.; Lubbers, E.R.; Murphy, N.P.; Little, S.C.; Musa, H.; Smith, S.A.; Unudurthi, S.D.; Rajaram, M.V.S.; Janssen, P.M.L.; et al. Novel Mechanistic Roles for Ankyrin-G in Cardiac Remodeling and Heart Failure. *Jacc Basic Transl. Sci.* **2018**, *3*, 675–689. [CrossRef]

29. Iqbal, Z.; Vandeweyer, G.; van der Voet, M.; Waryah, A.M.; Zahoor, M.Y.; Besseling, J.A.; Roca, L.T.; Vulto-van Silfhout, A.T.; Nijhof, B.; Kramer, J.M.; et al. Homozygous and heterozygous disruptions of ANK3: At the crossroads of neurodevelopmental and psychiatric disorders. *Hum. Mol. Genet.* **2013**, *22*, 1960–1970. [CrossRef]

30. Bennett, V.; Stenbuck, P.J. The membrane attachment protein for spectrin is associated with band 3 in human erythrocyte membranes. *Nature* **1979**, *280*, 468–473. [CrossRef]

31. Narla, J.; Mohandas, N. Red cell membrane disorders. *Int. J. Lab. Hematol.* **2017**, *39*, 47–52. [CrossRef] [PubMed]

32. Satchwell, T.J.; Bell, A.J.; Hawley, B.R.; Pellegrin, S.; Mordue, K.E.; van Deursen, C.T.B.M.; Braak, N.H.-T.; Huls, G.; Leers, M.P.G.; Overwater, E.; et al. Severe Ankyrin-R deficiency results in impaired surface retention and lysosomal degradation of RhAG in human erythroblasts. *Haematologica* **2016**, *101*, 1018–1027. [CrossRef] [PubMed]

33. Bennett, V.; Baines, A.J. Spectrin and ankyrin-based pathways: Metazoan inventions for integrating cells into tissues. *Physiol. Rev.* **2001**, *81*, 1353–1392. [CrossRef] [PubMed]

34. Curran, J.; Mohler, P.J. Alternative paradigms for ion channelopathies: Disorders of ion channel membrane trafficking and posttranslational modification. *Annu. Rev. Physiol.* **2015**, *77*, 505–524. [CrossRef] [PubMed]

35. Jenkins, P.M.; Kim, N.; Jones, S.L.; Tseng, W.C.; Svitkina, T.M.; Yin, H.H.; Bennett, V. Giant ankyrin-G: A critical innovation in vertebrate evolution of fast and integrated neuronal signaling. *Proc. Natl. Acad. Sci. USA* **2015**, *112*, 957–964. [CrossRef] [PubMed]

36. Camors, E.; Mohler, P.J.; Bers, D.M.; Despa, S. Ankyrin-B reduction enhances Ca spark-mediated SR Ca release promoting cardiac myocyte arrhythmic activity. *J. Mol. Cell Cardiol.* **2012**, *52*, 1240–1248. [CrossRef] [PubMed]

37. Kashef, F.; Li, J.; Wright, P.; Snyder, J.; Suliman, F.; Kilic, A.; Higgins, R.S.D.; Anderson, M.E.; Binkley, P.F.; Hund, T.J.; et al. Ankyrin-B protein in heart failure: Identification of a new component of metazoan cardioprotection. *J. Biol. Chem.* **2012**, *287*, 30268–30281. [CrossRef]

38. He, M.; Tseng, W.-C.; Bennett, V. A single divergent exon inhibits ankyrin-B association with the plasma membrane. *J. Biol. Chem.* **2013**, *288*, 14769–14779. [CrossRef]

39. Yang, R.; Walder-Christensen, K.K.; Kim, N.; Wu, D.; Lorenzo, D.N.; Badea, A.; Jiang, Y.-H.; Yin, H.H.; Wetsel, W.C.; Bennett, V. ANK2 autism mutation targeting giant ankyrin-B promotes axon branching and ectopic connectivity. *Proc. Natl. Acad. Sci. USA* **2019**, *116*, 15262–15271. [CrossRef]

40. Koenig, S.N.; Mohler, P.J. The evolving role of ankyrin-B in cardiovascular disease. *Heart Rhythm* **2017**, *14*, 1884–1889. [CrossRef]

41. El Refaey, M.M.; Mohler, P.J. Ankyrins and Spectrins in Cardiovascular Biology and Disease. *Front. Physiol.* **2017**, *8*, 852. [CrossRef] [PubMed]

42. Chu, L.; Greenstein, J.L.; Winslow, R.L. Na$^+$ microdomains and sparks: Role in cardiac excitation-contraction coupling and arrhythmias in ankyrin-B deficiency. *J. Mol. Cell Cardiol.* **2019**, *128*, 145–157. [CrossRef] [PubMed]

43. Michaely, P.; Tomchick, D.R.; Machius, M.; Anderson, R.G.W. Crystal structure of a 12 ANK repeat stack from human ankyrinR. *Embo. J.* **2002**, *21*, 6387–6396. [CrossRef] [PubMed]

44. Wang, C.; Yu, C.; Ye, F.; Wei, Z.; Zhang, M. Structure of the ZU5-ZU5-UPA-DD tandem of ankyrin-B reveals interaction surfaces necessary for ankyrin function. *Proc. Natl. Acad. Sci. USA* **2012**, *109*, 4822–4827. [CrossRef]

45. Zhu, W.; Wang, C.; Hu, J.; Wan, R.; Yu, J.; Xie, J.; Ma, J.; Guo, L.; Ge, J.; Qiu, Y.; et al. Ankyrin-B Q1283H Variant Linked to Arrhythmias Via Loss of Local Protein Phosphatase 2A Activity Causes Ryanodine Receptor Hyperphosphorylation. *Circulation* **2018**, *138*, 2682–2697. [CrossRef]

46. Qu, F.; Lorenzo, D.N.; King, S.J.; Brooks, R.; Bear, J.E.; Bennett, V. Ankyrin-B is a PI3P effector that promotes polarized α5β1-integrin recycling via recruiting RabGAP1L to early endosomes. *Elife* **2016**, *5*, e20417. [CrossRef]

47. Kontrogianni-Konstantopoulos, A.; Bloch, R.J. Obscurin: A multitasking muscle giant. *J. Muscle Res. Cell Motil.* **2005**, *26*, 419–426. [CrossRef]

48. Hall, T.G.; Bennett, V. Regulatory domains of erythrocyte ankyrin. *J. Biol. Chem.* **1987**, *262*, 10537–10545.

49. Davis, L.H.; Davis, J.Q.; Bennett, V. Ankyrin regulation: An alternatively spliced segment of the regulatory domain functions as an intramolecular modulator. *J. Biol. Chem.* **1992**, *267*, 18966–18972.

50. Lopes, L.R.; Syrris, P.; Guttmann, O.P.; O'Mahony, C.; Tang, H.C.; Dalageorgou, C.; Jenkins, S.; Hubank, M.; Monserrat, L.; McKenna, W.J.; et al. Novel genotype-phenotype associations demonstrated by high-throughput sequencing in patients with hypertrophic cardiomyopathy. *Heart* **2015**, *101*, 294–301. [CrossRef]

51. Ichikawa, M.; Aiba, T.; Ohno, S.; Shigemizu, D.; Ozawa, J.; Sonoda, K.; Fukuyama, M.; Itoh, H.; Miyamoto, Y.; Tsunoda, T.; et al. Phenotypic Variability of *ANK2* Mutations in Patients With Inherited Primary Arrhythmia Syndromes. *Circ. J.* **2016**, *80*, 2435–2442. [CrossRef] [PubMed]

52. Robaei, D.; Ford, T.; Ooi, S.-Y. Ankyrin-B Syndrome: A Case of Sinus Node Dysfunction, Atrial Fibrillation and Prolonged QT in a Young Adult. *Heartlung Circ.* **2015**, *24*, e31–e34. [CrossRef] [PubMed]

53. Schott, J.J.; Charpentier, F.; Peltier, S.; Foley, P.; Drouin, E.; Bouhour, J.B.; Donnelly, P.; Vergnaud, G.; Bachner, L.; Moisan, J.P. Mapping of a gene for long QT syndrome to chromosome 4q25-27. *Am. J. Hum. Genet.* **1995**, *57*, 1114–1122. [PubMed]

54. Swayne, L.A.; Murphy, N.P.; Asuri, S.; Chen, L.; Xu, X.; McIntosh, S.; Wang, C.; Lancione, P.J.; Roberts, J.D.; Kerr, C.; et al. Novel Variant in the ANK2 Membrane-Binding Domain Is Associated With Ankyrin-B Syndrome and Structural Heart Disease in a First Nations Population With a High Rate of Long QT Syndrome. *Circ. Cardiovasc. Genet.* **2017**, *10*, e001537. [CrossRef]

55. Huq, A.J.; Pertile, M.D.; Davis, A.M.; Landon, H.; James, P.A.; Kline, C.F.; Vohra, J.; Mohler, P.J.; Delatycki, M.B. A Novel Mechanism for Human Cardiac Ankyrin-B Syndrome due to Reciprocal Chromosomal Translocation. *Heart Lung Circ.* **2017**, *26*, 612–618. [CrossRef] [PubMed]

56. Corrado, D.; Link, M.S.; Calkins, H. Arrhythmogenic Right Ventricular Cardiomyopathy. *N. Engl. J. Med.* **2017**, *376*, 61–72. [CrossRef]

Review

Trafficking of Stretch-Regulated TRPV2 and TRPV4 Channels Inferred Through Interactomics

Pau Doñate-Macián [1,2], Jennifer Enrich-Bengoa [1,3], Irene R. Dégano [4,5,6], David G. Quintana [1] and Alex Perálvarez-Marín [1,3,*]

[1] Biophysics Unit, Department of Biochemistry and Molecular Biology, School of Medicine, Universitat Autònoma de Barcelona, 08193 Cerdanyola del Vallés, Catalonia, Spain; pabdoama@gmail.com (P.D.-M.); jennifer.enrich@uab.cat (J.E.-B.); DavidG.Quintana@uab.cat (D.G.Q.)

[2] Laboratory of Molecular Physiology, Department of Experimental and Health Sciences, Pompeu Fabra University, 08003 Barcelona, Catalonia, Spain

[3] Institut de Neurociències, Universitat Autònoma de Barcelona, 08193 Cerdanyola del Vallés, Catalonia, Spain

[4] CIBER Cardiovascular Diseases (CIBERCV), Instituto de Salud Carlos III, 28029 Madrid, Spain; iroman@imim.es

[5] REGICOR Study Group, Cardiovascular Epidemiology and Genetics Group, IMIM (Hospital Del Mar Medical Research Institute), 08003 Barcelona, Catalonia, Spain

[6] Faculty of Medicine, University of Vic-Central University of Catalonia (UVic-UCC), 08500 Vic, Spain

* Correspondence: alex.peralvarez@uab.cat; Tel.: +34-93-581-4504

Received: 14 October 2019; Accepted: 25 November 2019; Published: 27 November 2019

Abstract: Transient receptor potential cation channels are emerging as important physiological and therapeutic targets. Within the vanilloid subfamily, transient receptor potential vanilloid 2 (TRPV2) and 4 (TRPV4) are osmo- and mechanosensors becoming critical determinants in cell structure and activity. However, knowledge is scarce regarding how TRPV2 and TRPV4 are trafficked to the plasma membrane or specific organelles to undergo quality controls through processes such as biosynthesis, anterograde/retrograde trafficking, and recycling. This review lists and reviews a subset of protein–protein interactions from the TRPV2 and TRPV4 interactomes, which is related to trafficking processes such as lipid metabolism, phosphoinositide signaling, vesicle-mediated transport, and synaptic-related exocytosis. Identifying the protein and lipid players involved in trafficking will improve the knowledge on how these stretch-related channels reach specific cellular compartments.

Keywords: ion channel trafficking; transient receptor potential channels; TRPV2; TRPV4; phosphatidylinositol signaling; stretch-related channels

---

1. Introduction

Transient receptor potential (TRP) channels are polymodal cation channels involved in somatosensation at the cellular and tissue levels in vertebrates [1]. Channels in the TRP family are in charge of sensing physical stimuli such as temperature or mechanical changes to trigger cation-mediated cell signal transduction pathways [2]. The vanilloid subfamily (TRPV) has six members (TRPV1–6), where TRPV1–4 have been long related to thermal sensing [3]. More recently, they have also been linked to mechanical stress, especially the TRPV2 and TRPV4 channels [4,5]. Accordingly, the function of TRPV2 and TRPV4 is expected to be essential in tissues with high mechanical shearing, such as skeletal and cardiac muscle.

Expression of TRPV2 is wide (Figure 1a) and TRPV2 is involved in several physiological processes [6], but the particular role of this channel in skeletal and cardiac muscle is gaining much attention [7–12]. The expression of TRPV4 is as ubiquitous as TRPV2 (Figure 1a), but it is prominent in epithelial tissues, evoking calcium currents in response to extracellular stimuli, such as temperature, osmotic changes, and mechanical stretch [13–15]. Mutations of TRPV4 are related to involved in oligomerization, trafficking, and degradation can result in genetic disorders like Brachyolmia, Charcot–Marie–Tooth disease type 2C, spinal muscular spinal muscular atrophy, arthrogryposis, and hereditary motor and sensory neuropathy type 2 [16,17]. The identification of these mutations is a first step to determine the pathogenesis of the associated diseases and to design specific therapies. In addition, the disruption of the folding-sensitive region of TRPV4 could be a therapeutic option for diseases in which TRPV4 increases its activity such as pain and skeletal dysplasias [18]. Recently, it has also been shown that TRPV4 affects the calcium balance in cardiomyocytes, affecting contractility, leading to cardiac tissue damage in heart physiopathology [19].

An important question regarding TRPV2 and TRPV4 is how these channels are trafficked to and recycled from the membrane. This is an essential aspect in TRPV2 and TRPV4 function as Ca^{2+}-dependent stretch-modulated channels. It has been shown that trafficking and/or translocation is a highly regulated process, and it may be dependent on channel activity. Translocation of TRPV2 to the membrane is driven by growth factors or chemotactic peptides [6], although is not clear yet whether TRPV2 is functional when at the plasma membrane and/or internal organelles, mainly because most the literature regarding TRPV2 trafficking is based on poor detection antibodies against TRPV2 [20]. The function of TRPV4 is exerted mainly at the plasma membrane and TRPV4 trafficking to the membrane is regulated by activators, such as GSK1016790A [18]. Controlled and regulated trafficking of ion channels, especially in excitatory tissues, is fundamental because of the possibility of cation leakage during trafficking, which leads to unbalanced cation homeostasis promoting cell toxicity and/or excitotoxicity. This review of the published protein–protein interactions for the TRPV2 and TRPV4 channels [21–24] intends to shed light about the proteins involved in the regulated and constitutive trafficking of these two mechanosensory cation channels.

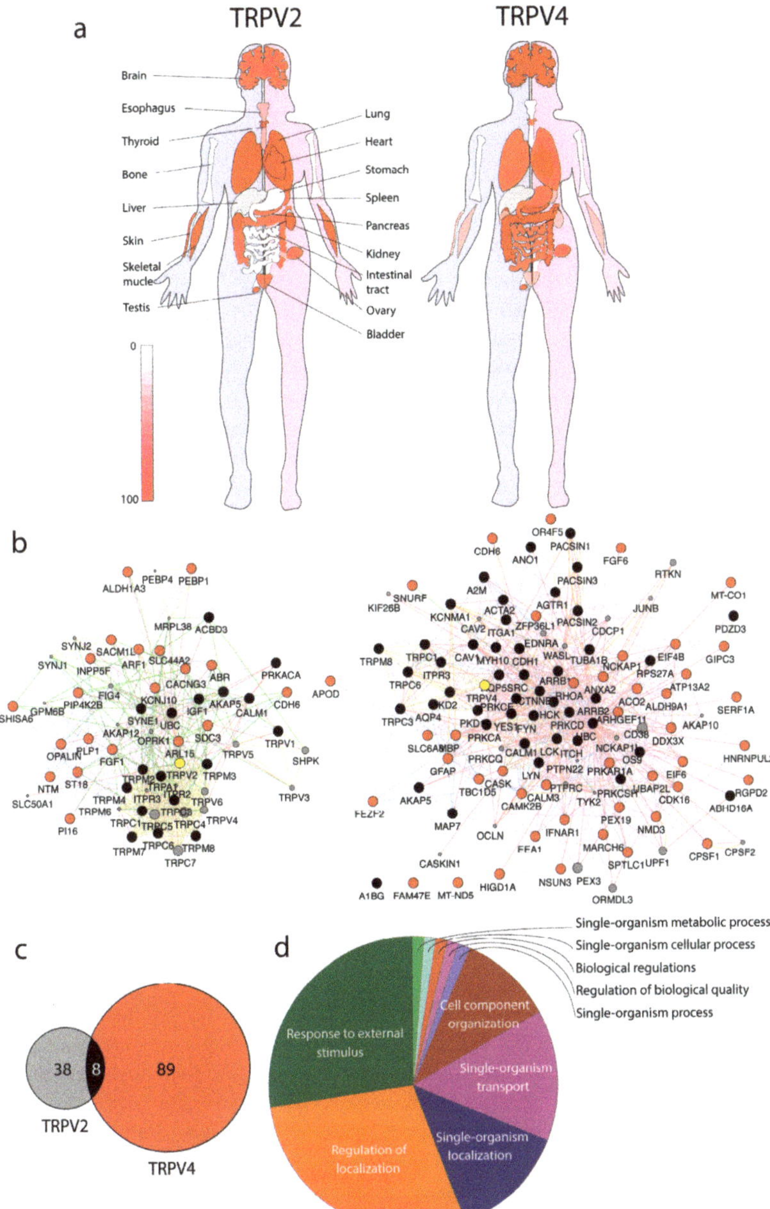

Figure 1. Transient receptor potential vanilloid subfamily TRPV2 and TRPV4 expression and interactomic profiles. (**a**) TRPV2 and TRPV4 tissue expression in humans; (**b**) TRPV2 (left) and TRPV4 (right) interactomes crossing published results (see main text for details). (**c**) Protein interactions overlap among TRPV2 and TRPV4 interactomes. (**d**) Main biological process terms defined by TRPV2 and TRPV4 gene set enrichment analysis (GSEA).

2. Sequence and Structure Determinants in Channel Trafficking

Beyond TRP biogenesis and oligomerization as tetramers [25], TRP channel trafficking is a complex mechanism, integrating processes such as membrane insertion, glycosylation, Golgi maturation, vesicle trafficking, and protein–protein interaction (PPI) [26]. The balance between these processes results in a correct protein distribution in the plasma membrane or the corresponding membrane compartment, leading to up- and down-regulation of the protein function. The trafficking structural and molecular determinants in TRP sequence are mediators of PPI and/or lipid–protein interactions (LPIs). In TRPV channels, both the N- and C-termini were involved in trafficking. The distal N-terminus of TRPV channels is highly variable and structurally disordered and likely to host several phosphorylation sites and PPI and LPI domains. Deletion of the distal N-terminus of TRPV2 is enough to deplete the channel trafficking to the plasma membrane [23], which has also been observed for TRPV5 [27]. In addition, the N-terminus of TRPV4 interacts with OS-9 in the endoplasmic reticulum (ER), preventing channel trafficking to the plasma membrane [28]. Protein kinase C and casein kinase substrate in neurins proteins (PACSIN) also bind to the TRPV4 distal N-terminal domain (interaction exclusive for TRPV4 among TRPV channels), enhancing the relative amount of TRPV4 in the plasma membrane [29]. As for the C-terminus, AKAP79/150 binding has been described [24], as well as the ankyrin repeat domain (ARD), which is involved in PPI, but also in the complex mechanism of trafficking of TRPV channels. The TRPV4 C-terminus is also involved in trafficking, as shown by C-terminus deletion mutants, resulting in TRPV4 accumulation in the ER [30]. The characteristic ankyrin repeat domain (ARD) is involved in PPI, but also in the complex mechanism of trafficking of TRPV channels. Experiments carried out to map in vitro TRPV2 topology [23] were performed with constructs lacking either most of the N-terminal domain of the channel (the first 74 amino acids ΔN74-TRPV2) or the first 336 amino acids at the distal N-terminus (corresponding to the ARD, ΔARD-TRPV2). Despite the N-terminal truncation, TRPV2 was properly folded within the lipid bilayer. The ΔARD-TRPV2 mutant was not able to traffic to the plasma membrane, as determined by confocal imaging and biotinylation assay [23]. Such a fact indicates that the N-terminal region may be needed for additional channel processes other than insertion in the ER membranes, such as channel tetramerization, glycosylation, or interaction with chaperone proteins to allow membrane translocation. Although it has not been described for TRPV2 [31], the ARD indeed plays a key role in channel oligomerization for TRPV4, as has been shown for ARD mutants in TRPV4 [32]. TRPV mutants lacking ARD or carrying point mutations in N- or C-terminal domains produce channels that seem unable to tetramerize, which emphasizes the need for the TRPV cytosolic domains to promote the correct oligomerization of the subunits [32,33].

Turnover of TRPV channels in the plasma membrane is also controlled by regulated exocytosis, a process mediated by phosphorylation and interaction with SNAP (soluble N-ethylmaleimide sensitive fusion attachment protein) receptor (SNARE) complex proteins. SNAREs are a protein complex of more than 60 members in mammalian cells that mediate vesicle fusion with their target membrane bound compartment. The complex of SNAREs has a relevant role in cellular processes such as neurotransmitter release in synapses, exocytosis, or autophagy [34,35]. Upon protein kinase C (PKC) activation, TRPV1 is recruited to the plasma membrane by SNARE mediated vesicle transport, leading to a potentiation of TRPV1 currents [36]. Interaction of TRPV1 and TRPV2 with Snapin and SynaptotagminIX (SYT9) [36,37]. For TRPV2, the interaction with SNAREs protein is mapped into the highly conserved region of the membrane proximal domain (MPD), pointing to the conservation of these interactions along the TRPV1–4 subfamily [37].

Regarding LPI, TRPV channels share highly conserved PIP2 binding domains [24]. Binding of PIP2 to TRPV1 [38], TRPV2 [39], and TRPV3 [40] was mapped to the highly conserved proximal C-terminal domain, contiguous to the TRP box. The TRPV4 PIP2 binding domain is pinpointed using bioinformatics [24], although TRPV4 has an additional PIP2 binding site in the distal N-terminal domain [41]. Two recent works suggest that phosphatidic acid (PA) mediates LPIs between the membrane proximal domain (pre-S1 domain) and the C-terminus of TRPV channels in the trafficking of TRPV channels to the plasma membrane [37,42].

3. Interactomics

Proteomic studies provide useful tools to understand the molecular mechanism of TRPV channels. Guilt-by-association approaches aiming for mid-throughput interactome discovery for TRPV2 and TRPV4 were used, combined with standard techniques such as co-immunoprecipitation in mammalian cell lines and a novel membrane-specific yeast two-hybrid (MYTH) methodology [21,22].

The MYTH is a split-ubiquitin-based system that allows the location of the bait protein (TRPV2, TRPV4) embedded in the membrane. First, a yeast strain is generated that constitutively expresses the ion channel fused to the C-terminus of ubiquitin followed by the transcription factor LexA-VP16 (TF). The bait strain is then transformed with a cDNA library containing more than 10^6 protein preys fused to the N-terminus of ubiquitin. The tag carries a mutation to hinder the spontaneous refolding of ubiquitin, which thus will only occur when a prey protein interacts with the bait. Once the interaction occurs and ubiquitin is refolded, the bait complex is cleaved and the TF is released. The traffic of TF to the nucleus starts the expression of reporter genes integrated in the yeast strain, allowing cells to grow in selective media lacking either Histidine (His) or Adenine (Ade). Moreover, the LacZ reporter gene is activated upon interaction, so that colonies carrying a putative interactor become blue stained in the presence of X-Gal. The blue color intensity may be used to determine whether interactions are strong or constitutive (intense blue color staining) or transient (pale blue color intensity) [43]. Using MYTH, 20 and 44 new interactors were found for TRPV2 and TRPV4, respectively [21,22] (Table S1 and Figure 1b,c).

Available PPIs for TRP channels are listed in the TRIP database [44]. For TRPV2 and TRPV4 PPI, refer to Supporting Table S1 and Figure 1b,c. The TRIP database is an excellent tool to get relevant information of biological processes regarding TRP channels. To the best of our knowledge, the database was last updated in August 2015, not including any new PPIs, such as the TRPV2 and TRPV4 membrane yeast two-hybrid (MYTH) dataset from these studies [21,24]. Figure 1 and Supplementary Table S1 list all the interactions for TRPV2 and TRPV4. However, for the PPI trafficking analysis performed in this study, only the PPIs identified in MYTH studies [21,22] are used. Using the list of TRPV2 and TRPV4 PPIs, a gene set enrichment analysis (GSEA) is performed, and then all ion channels and transporters filtered out to avoid the bias towards transport and cation transport gene ontology terms. Out of the 59 protein list, the GSEA resulted in 10 enriched biological process categories, including the following: neural nucleus development (GO:0048857); phosphatidylinositol biosynthetic process (GO:0006661); myelin sheath (GO:0043209); phospholipid biosynthetic process (GO:0008654); regulation of cellular amide metabolic process (GO:0034248); modulation of chemical synaptic transmission (GO:0050804); regulation of vesicle-mediated transport (GO:0060627); Ras GTPase binding (GO:0017016); regulation of cellular localization (GO:0060341); and cellular lipid metabolic process (GO:0044255).

4. Understanding Channel Trafficking through Protein–Protein Interactions

In this review, the focus lies in terms related to two aspects: (i) lipid and phosphoinositides, and (ii) synaptic and vesicle-regulated trafficking. Because both aspects are tightly entangled, they and are difficult to study independently [21–24,37]. Lipids are key effectors in signal transduction and protein function, but also in protein trafficking [45]. Phosphoinositides (PIs) provide cellular organelles with specific lipid signatures, facilitating the binding of specific proteins. Membrane protein traffic is mediated by accessory proteins containing distinct lipid–protein binding domains, which promote binding depending on the lipids physico-chemical nature [45]. Regulated (e.g., synaptic-based exocytosis) and constitutive vesicle-mediated transport for TRP channels have already been reviewed, mostly based on TRPV1 studies [46]. This review aims to expand the knowledge on vesicle/synaptic-based trafficking by shedding some light on accessory proteins involved in these processes, as listed in Table 1.

Table 1. List of interactors related to trafficking processes derived from a membrane yeast two hybrid approach.

Gene Symbol	Interactor	Gene Name	Gene ID	Process [a]
ABHD16A	TRPV4	abhydrolase domain containing 16A	7920	1
ANXA2	TRPV4	annexin A2	302	1, 2
ARF1	TRPV2	Adenosine diphosphate (ADP) ribosylation factor 1	375	1, 2, 3, 4, 5
ATP13A2	TRPV4	Adenosine triphosphate (ATP)ase cation transporting 13A2	23400	2,3
CAMK2B	TRPV4	calcium/calmodulin dependent protein kinase II beta	816	4
CASK	TRPV4	calcium/calmodulin dependent serine protein kinase	8573	2,3,4
CDK16	TRPV4	cyclin dependent kinase 16	5127	2
INPP5F	TRPV2	inositol polyphosphate-5-phosphatase F	22876	1, 2, 3, 5, 6
PIP4K2B	TRPV2	phosphatidylinositol-5-phosphate 4-kinase type 2 beta	8396	1, 5, 6
SACM1L	TRPV2	SAC1 like phosphatidylinositide phosphatase	22908	1, 5, 6
SDC3	TRPV2	syndecan 3	9672	1, 2, 3, 4, 5
SHISA6	TRPV2	shisa family member 6	388336	2, 4
SNAPIN	TRPV2/TRPV1	SNAP associated protein	23557	2, 3, 4
SPTLC1	TRPV4	serine palmitoyltransferase long chain base subunit 1	10558	1, 6
SYT9	TRPV2/TRPV1	synaptotagmin 9	143425	2, 3, 4
TBC1D5	TRPV4	TBC1 domain family member 5	9779	3

[a] Processes: 1, cellular lipid metabolism; 2, regulation of cell localization; 3, regulation of vesicle-mediated transport; 4, modulation of chemical synaptic transmission; 5, phosphatidylinositol biosynthesis; 6, phospholipid biosynthesis. TRPV, transient receptor potential vanilloid subfamily; SNAP, soluble NSF attachment protein.

Among the list of interactors (Table S1), the literature is revised regarding proteins/enzymes related to phosphoinositide signaling as markers of specific lipidic composition among cellular organelles, but also to trafficking and vesicle-mediated accessory proteins (Figure 2). This review provides an overview of the TRPV2 and TRPV4 trafficking process. Side information on TRPV2 and TRPV4 trafficking is found in the literature, such as the role of PACSIN3 for TRPV4 [29,41], recombinase gene activator (RGA) protein, and ras-related protein 7 (Rab7) for TRPV2 [47,48], and the Snapin/Syt9 pair for TRPV1 and TRPV2 [21,36,37]. However, the role of the lipid-mediated PPI for TRPV channels deserves extra attention, not only because of the complexity of the mechanism, but also because of the diversity of mechanisms depending on the tissue of interest (Figure 1a).

The lipase/acyl-transferase dehydrogenase enzyme (ABHD16A) is a TRPV4 interactor involved in lipid metabolism [49]. Proteins of the ABHD family are involved in lipidic modifications, such as palmitoylation, and in negative regulation of α-amino-3-hydroxy-5-methyl-4-isoxazolepropionic acid (AMPA) receptor trafficking. The depalmitoylase ABHD17A is crucial for synaptic targeting and vesicle sorting of AMPA receptors, through PSD-95 depalmitoylation [49,50]. In line with AMPA, but also to *N*-methyl-D-aspartate (NMDA) receptors trafficking, the calcium/calmodulin dependent protein kinase II beta (CAMK2B) and the calcium/calmodulin dependent serine protein kinase (CASK), both TRPV4 interactors, are trafficking regulatory proteins [51–53]. Another kinase interacting with TRPV4 is the cyclin dependent kinase 16 (CDK16), a key regulator of vesicle trafficking [54–56]. The kinase CDK16 interacts directly with COPII complexes modulating secretory cargo transport.

Annexin 2 (AnxA2) is a lipid raft associated trafficking factor in the plasma membrane and the endosomal system, related to both endo- and exocytosis [57]. Binding of AnxaA2 to phospholipids in a Ca^{2+}-dependent manner [58], generating microdomains suitable for the binding of membrane proteins, such as the renal cotransporter NKCC2 [59]. AnxA2 has been shown to interact with fibroblast growth factor 1 (FGF1) (Table S1), forming heteroligomers capable of interacting with acidic membrane lipids, such as PA. Viral infection hijacks AnxA2, which is used for the virus advantage in cell-attachment, replication, and proliferation processes [60]; thus, AnxA2 is a likely candidate protein to play a role in the TRPV4-DDX3X mechanism in viral infectivity [22].

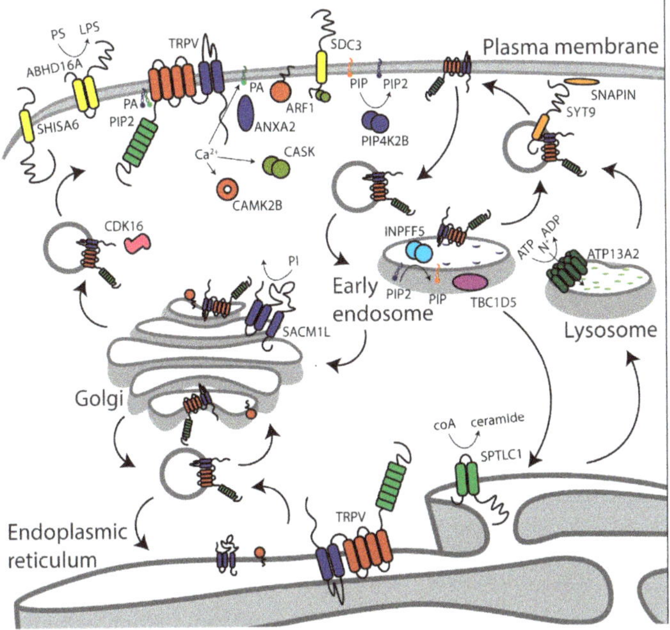

Figure 2. Cellular overview of the main TRPV2 and TRPV4 trafficking-related protein–protein interactions (PPIs) organized by cellular compartment (see main text for details).

The ADP-ribosylation factor (Arf) small G proteins, such as Arf1, are related to lipid droplet metabolism, clathrin independent endocytosis, and other membrane dynamics processes [61–63]. N-terminus miristoylated Arf1 only tethers to the membrane when bound to guanosine triphosphate (GTP) [64]. Guanosine exchange factors (GEF) and GTPase activating proteins (GAP) are required by Arfs. Hydrolysis of GTP by Arf and Arf-like proteins regulate the enzymatic activity of proteins, such as PI kinases and phosphatases. Among the TRPV2 interactors, Phosphatidylinositol-5-phosphate 4-kinase type 2 beta (PIP4K2B), inositol polyphosphate-5-phosphatase F (INPP5F, also known as Sac2), and SAC1 like phosphatidylinositide phosphatase (SACM1L, also known as Sac1) are enzymes involved in phosphoinositide regulation as signals for membrane traffic. The Arf1 pathway is related to phospholipase D (PLD), which is responsible for the production of phosphatidic acid as a signaling molecule, shown to interact with TRPV channels [37,42]. The Arf1–PLD pathways are responsible for vesicle-mediated endocytosis and exocytosis, as well as the formation of multivesicular bodies (MVBs) through phosphoinositides binding/signaling [45]. The putative interaction between TRPV2 and Arf1 is an important hint regarding vesicle-mediated constitutive trafficking of TRPV channels. Other proteins that could be related to the Arf1 pathway, identified as TRPV2 and TRPV4 interactions, are Arl15 (Arf-like protein) and the CALM/CamK2B/Cask subset, respectively (Table S1 and Table 1).

Enzymes involved in phosphoinositide regulation are TRPV2/TRPV4 interactors, such as PIP4K2B and the SACM1L and INPP5F pair, Sac1 and Sac2, respectively. The kinase PIP4K2B regulates the levels of phosphatidylinositol 5-phosphate (PI5P) [65] by converting it to phosphatidylinositol 4,5-biphosphate (PI(4,5)P2), a lipid involved in TRPV2 and TRPV4 function [6,41]. The non-abundant PI5P phosphoinositide is related to the Akt kinase pathway and relevant for several cellular processes, such as survival and cell growth, with a prominent role in cancer [66]. Hydrolysis of phosphatidylinositol 3-phosphate (PI3P), phosphatidylinositol 4-phosphate (PI4P), and phosphatidylinositol 3,5-bisphosphate(PI(3,5)P2) is carried out by SACM1L/Sac1, which is enriched at the Golgi membrane, but is also present in ER membranes [67]. The preferential substrate for this enzyme is PI4P. The spatial distribution of

PI4P in Golgi is defined by SACM1L/Sac1, creating the optimal conditions to maintain the cisternal identity of the Golgi, which is critical to membrane protein trafficking [68]. Inositol polyphosphate-5-phosphatase F (INPP5F, also known as Sac2) has a specific activity for PI(4,5)P2 and phosphatidylinositol 3,4,5-trisphosphate (PI(3,4,5)P3) to generate PI4P and PI(3,4)P2. INPP5F/Sac2 colocalizes with early endosomal markers and is related to the regulation of endocytic recycling [69].

The ATPase cation transporting 13A2 (ATP13A2 or PARK9) is an endo-/lysosomal associated ATPase related to intracellular trafficking under proteotoxic stress [70]. In a MYTH-based proteomics approach [71], the authors have identified 43 PARK9-interacting proteins related to trafficking. The putative TRPV4-PARK9 and the TRPV2-INPP5F interactions identified in studies [21,22] could be interesting to understand TRPV channels recycling in the endosome/lysosome pathways.

Syndecan-3 (SDC3) is a heparan-sulfate proteoglycan and a TRPV2 interactor, related to leukocyte migration [72]. Modification of cytoskeleton is carried out by SDC3, but no relationship between SDC3 and membrane protein trafficking has been shown so far. However, SDC4 modulates the activity and membrane expression of TRPC6 in glomerular permeabilization [73].

The trafficking of AMPA receptors towards the synapse has been thoroughly studied [74], and among the proteins involved in locating AMPA receptors at the synapse, SHISA6 is found as a TRPV2 interactor. The C-terminal characteristic PDZ domain in SHISA6 binds to post-synaptic density protein 95 (PSD95), which confines AMPA receptors at the postsynaptic density [75]. Proteins, such as SHISA6, ABHD16A, Snapin, and SYT9 (Snapin and SYT9 revised earlier), interact with TRPV channels, mediating synaptic exocytosis, suggesting conservation of some elements in regulated exocytosis of TRPV channels [21,36,37,46].

The TRPV4 interactor serine palmitoyltransferase long chain base subunit 1 (SPTLC1), which resides in the ER, drives the synthesis of sphingomyelin [76], sphingolipid that needs to be trafficked to the plasma membrane. Specific lipid composition in membrane domains may argue for the need of specific lipids bound to TRPV channels during trafficking. SPTLC1, teaming up with ORMDL3, are involved in ceramide synthesis [77,78], in response to ER stress and calcium homeostasis, factors influencing the trafficking of membrane proteins, such as TRPV channels.

The TBC1 domain family member 5 (TBC1D5) plays a significant role in the regulation of the endosomal pathway. TBC1D5 interacts with TRPV4 in a yeast two-hybrid approach [22]. In the endosomal system, TBC1D5 inhibits the retromer and promoting autophagy. It is also a key factor in membrane turnover and membrane protein recycling through Rab7 [79]. The Rab7 and TRPV2 pair has been shown to colocalize in the endosome. As a result of such a trafficking situation, TRPV2 relates to nervous system development by enhancing neurite outgrowth [48]. Among TRPV channels, TRPV2 is the most likely to play a role in the regulation of the endosomal pathway [80].

5. Concluding Remarks

This review of the literature aims toward a better understanding of the mechanisms driving the trafficking of TRPV2 and TRPV4 to specific membrane compartments, derived from a guilt-by-association approach based on PPI (Table S1). The interaction of TRPV2 and TRPV4 with lipids and/or with lipid modifying enzymes points to the fact that the lipid environment is fundamental for TRPV2 and TRPV4 trafficking, beyond the TRPV2 and TRPV4 lipid requirement for ion channel function. Future perspectives should include the study of the tight inactivation mechanisms of ion channels involved in the trafficking of these proteins, to prevent cation leakage inside the cell. In the case of TRPV2 and TRPV4 (and other TRP channels), the focus so far has resided on which lipid signaling molecules activate/inhibit the channel at the site of function. The identification of LPI and PPI interactions preventing cation leakage during trafficking toward avoiding detrimental cell toxicity effects provides another potentially interesting area of study, which may be especially relevant in tissues under high stretch stress conditions, such as skeletal and cardiac muscle. Knowledge on how TRPV2 and TRPV4 are specifically trafficked in these tissues might provide invaluable benefits in the therapeutic management of muscle physiopathology, where cation transport and balance play a cardinal role.

Supplementary Materials: The following are available online at http://www.mdpi.com/2218-273X/9/12/791/s1, Table S1. TRPV2 and TRPV4 interactomes.

Author Contributions: Conceptualization P.D.-M., J.E.-B., I.R.D., D.G.Q. and A.P.-M.; writing—original draft preparation, P.D.-M. and A.P.-M.; writing—review and editing, P.D.-M., J.E.-B., I.R.D., D.G.Q. and A.P.-M.

Funding: This work was supported by Marie Curie International Outgoing Fellowship within the 7th European Community Framework Programme (PIOF-GA-2009-237120 to A.P.-M.), a Universitat Autònoma de Barcelona-Programa Banco de Santander Fellowship, and the Spanish Government grant MINECO BFU2017-87843-R to A.P.-M. P.D.-M. was a recipient of an FI fellowship from Generalitat de Catalunya (FI-2013FIB00251). I.R.D. was supported by the Carlos III Health Institute, co-financed with European Union *European Regional Development Funds* (CIBERCV CB16/11/00229), and by the Strategic Plan for research and health innovation (PERIS) (SLT006/17/00029).

Conflicts of Interest: The authors declare no conflict of interest. The funders had no role in the design of the study; in the collection, analyses, or interpretation of data; in the writing of the manuscript; or in the decision to publish the results.

References

1. Nilius, B.; Owsianik, G. The transient receptor potential family of ion channels. *Genome Biol.* **2011**, *12*, 218. [CrossRef]

2. Ramsey, I.S.; Delling, M.; Clapham, D.E. AN INTRODUCTION TO TRP CHANNELS. *Annu. Rev. Physiol.* **2006**, *68*, 619–647. [CrossRef] [PubMed]

3. Gunthorpe, M.J.; Benham, C.D.; Randall, A.; Davis, J.B. The diversity in the vanilloid (TRPV) receptor family of ion channels. *Trends Pharmacol. Sci.* **2002**, *23*, 183–191. [CrossRef]

4. Shibasaki, K. Physiological significance of TRPV2 as a mechanosensor, thermosensor and lipid sensor. *J. Physiol. Sci.* **2016**, *66*, 359–365. [CrossRef] [PubMed]

5. Ho, T.C.; Horn, N.A.; Huynh, T.; Kelava, L.; Lansman, J.B. Evidence TRPV4 contributes to mechanosensitive ion channels in mouse skeletal muscle fibers. *Channels* **2012**, *6*, 246–254. [CrossRef]

6. Perálvarez-Marín, A.; Doñate-Macian, P.; Gaudet, R. What do we know about the transient receptor potential vanilloid 2 (TRPV2) ion channel? *FEBS J.* **2013**, *280*, 5471–5487. [CrossRef]

7. Robbins, N.; Koch, S.E.; Rubinstein, J. Targeting TRPV1 and TRPV2 for potential therapeutic interventions in cardiovascular disease. *Transl. Res.* **2013**, *161*, 469–476. [CrossRef]

8. Sabourin, J.; Cognard, C.; Constantin, B. Regulation by scaffolding proteins of canonical transient receptor potential channels in striated muscle. *J. Muscle Res. Cell Motil.* **2009**, *30*, 289–297. [CrossRef]

9. Watanabe, H.; Murakami, M.; Ohba, T.; Ono, K.; Ito, H. The pathological role of transient receptor potential channels in heart disease. *Circ. J.* **2009**, *73*, 419–427. [CrossRef]

10. Lorin, C.; Vögeli, I.; Niggli, E. Dystrophic cardiomyopathy: role of TRPV2 channels in stretch-induced cell damage. *Cardiovasc. Res.* **2015**, *106*, 153–162. [CrossRef]

11. Jones, S.; Mann, A.; Worley, M.C.; Fulford, L.; Hall, D.; Karani, R.; Jiang, M.; Robbins, N.; Rubinstein, J.; Koch, S.E. The role of transient receptor potential vanilloid 2 channel in cardiac aging. *Aging Clin. Exp. Res.* **2017**, *29*, 863–873. [CrossRef] [PubMed]

12. Aguettaz, E.; Bois, P.; Cognard, C.; Sebille, S. Stretch-activated TRPV2 channels: Role in mediating cardiopathies. *Prog. Biophys. Mol. Biol.* **2017**, *130*, 273–280. [CrossRef] [PubMed]

13. Güler, A.D.; Lee, H.; Iida, T.; Shimizu, I.; Tominaga, M.; Caterina, M. Heat-evoked activation of the ion channel, TRPV4. *J. Neurosci.* **2002**, *22*, 6408–6414. [CrossRef] [PubMed]

14. Liedtke, W.; Choe, Y.; Martí-Renom, M.A.; Bell, A.M.; Denis, C.S.; AndrejŠali; Hudspeth, A.J.; Friedman, J.M.; Heller, S. Vanilloid Receptor–Related Osmotically Activated Channel (VR-OAC), a Candidate Vertebrate Osmoreceptor. *Cell* **2000**, *103*, 525–535. [CrossRef]

15. Suzuki, M.; Mizuno, A.; Kodaira, K.; Imai, M. Impaired pressure sensation in mice lacking TRPV4. *J. Biol. Chem.* **2003**, *278*, 22664–22668. [CrossRef] [PubMed]

16. Velilla, J.; Marchetti, M.M.; Toth-Petroczy, A.; Grosgogeat, C.; Bennett, A.H.; Carmichael, N.; Estrella, E.; Darras, B.T.; Frank, N.Y.; Krier, J.; et al. Homozygous TRPV4 mutation causes congenital distal spinal muscular atrophy and arthrogryposis. *Neurol. Genet.* **2019**, *5*, e312. [CrossRef]

17. Verma, P.; Kumar, A.; Goswami, C. TRPV4-mediated channelopathies. *Channels* **2010**, *4*, 319–328. [CrossRef]

18. Lei, L.; Cao, X.; Yang, F.; Shi, D.J.; Tang, Y.Q.; Zheng, J.; Wang, K. A TRPV4 channel C-terminal folding recognition domain critical for trafficking and function. *J. Biol. Chem.* **2013**, *288*, 10427–10439. [CrossRef]

19. Jones, J.L.; Peana, D.; Veteto, A.B.; Lambert, M.D.; Nourian, Z.; Karasseva, N.G.; Hill, M.A.; Lindman, B.R.; Baines, C.P.; Krenz, M.; et al. TRPV4 increases cardiomyocyte calcium cycling and contractility yet contributes to damage in the aged heart following hypoosmotic stress. *Cardiovasc. Res.* **2019**, *115*, 46–56. [CrossRef]

20. Cohen, M.R.; Huynh, K.W.; Cawley, D.; Moiseenkova-Bell, V.Y. Understanding the cellular function of TRPV2 channel through generation of specific monoclonal antibodies. *PLoS ONE* **2013**, *8*, e85392. [CrossRef]

21. Doñate-Macián, P.; Gómez, A.; Dégano, I.R.; Perálvarez-Marín, A. A TRPV2 interactome-based signature for prognosis in glioblastoma patients. *Oncotarget* **2018**, *9*, 18400–18409. [CrossRef] [PubMed]

22. Doñate-Macián, P.; Jungfleisch, J.; Pérez-Vilaró, G.; Rubio-Moscardo, F.; Perálvarez-Marín, A.; Diez, J.; Valverde, M.A. The TRPV4 channel links calcium influx to DDX3X activity and viral infectivity. *Nat. Commun.* **2018**, *9*, 1–13. [CrossRef] [PubMed]

23. Doñate-Macian, P.; Bañó-Polo, M.; Vazquez-Ibar, J.-L.; Mingarro, I.; Perálvarez-Marín, A. Molecular and topological membrane folding determinants of transient receptor potential vanilloid 2 channel. *Biochem. Biophys. Res. Commun.* **2015**, *462*, 221–226. [CrossRef] [PubMed]

24. Doñate-Macián, P.; Perálvarez-Marín, A. Dissecting domain-specific evolutionary pressure profiles of transient receptor potential vanilloid subfamily members 1 to 4. *PLoS ONE* **2014**, *9*, e110715. [CrossRef] [PubMed]

25. Huynh, K.W.; Cohen, M.R.; Jiang, J.; Samanta, A.; Lodowski, D.T.; Zhou, Z.H.; Moiseenkova-Bell, V.Y. Structure of the full-length TRPV2 channel by cryo-EM. *Nat Commun* **2016**, *7*, 11130. [CrossRef] [PubMed]

26. Cayouette, S.; Boulay, G. Intracellular trafficking of TRP channels. *Cell Calcium* **2007**, *42*, 225–232. [CrossRef]

27. De Groot, T.; Van Der Hagen, E.A.E.; Verkaart, S.; Te Boekhorst, V.A.M.; Bindels, R.J.M.; Hoenderop, J.G.J. Role of the transient receptor potential vanilloid 5 (TRPV5) protein N terminus in channel activity, tetramerization, and trafficking. *J. Biol. Chem.* **2011**, *286*, 32132–32139. [CrossRef]

28. Wang, Y.; Fu, X.; Gaiser, S.; Köttgen, M.; Kramer-Zucker, A.; Walz, G.; Wegierski, T. OS-9 regulates the transit and polyubiquitination of TRPV4 in the endoplasmic reticulum. *J. Biol. Chem.* **2007**, *282*, 36561–36570. [CrossRef]

29. Cuajungco, M.P.; Grimm, C.; Oshima, K.; D'Hoedt, D.; Nilius, B.; Mensenkamp, A.R.; Bindels, R.J.M.; Plomann, M.; Heller, S. PACSINs bind to the TRPV4 cation channel: PACSIN 3 modulates the subcellular localization of TRPV4. *J. Biol. Chem.* **2006**, *281*, 18753–18762. [CrossRef]

30. Becker, D.; Müller, M.; Leuner, K.; Jendrach, M. The C-terminal domain of TRPV4 is essential for plasma membrane localization. *Mol. Membr. Biol.* **2008**, *25*, 139–151. [CrossRef]

31. Jin, X.; Touhey, J.; Gaudet, R. Structure of the N-terminal ankyrin repeat domain of the TRPV2 ion channel. *J. Biol. Chem.* **2006**, *281*, 25006–25010. [CrossRef] [PubMed]

32. Arniges, M.; Fernández-Fernández, J.M.; Albrecht, N.; Schaefer, M.; Valverde, M.A. Human TRPV4 channel splice variants revealed a key role of ankyrin domains in multimerization and trafficking. *J. Biol. Chem.* **2006**, *281*, 1580–1586. [CrossRef] [PubMed]

33. Garcia-Elias, A.; Berna-Erro, A.; Rubio-Moscardo, F.; Pardo-Pastor, C.; Mrkonjić, S.; Sepúlveda, R.V.; Vicente, R.; González-Nilo, F.; Valverde, M.A. Interaction between the Linker, Pre-S1, and TRP Domains Determines Folding, Assembly, and Trafficking of TRPV Channels. *Structure* **2015**, 1–10. [CrossRef] [PubMed]

34. Ramakrishnan, N.A.; Drescher, M.J.; Drescher, D.G. The SNARE complex in neuronal and sensory cells. *Mol. Cell. Neurosci.* **2012**, *50*, 58–69. [CrossRef] [PubMed]

35. Moreau, K.; Ravikumar, B.; Renna, M.; Puri, C.; Rubinsztein, D.C. Autophagosome precursor maturation requires homotypic fusion. *Cell* **2011**, *146*, 303–317. [CrossRef] [PubMed]

36. Morenilla-Palao, C.; Planells-Cases, R.; García-Sanz, N.; Ferrer-Montiel, A. Regulated exocytosis contributes to protein kinase C potentiation of vanilloid receptor activity. *J. Biol. Chem.* **2004**, *279*, 25665–25672. [CrossRef]

37. Doñate-Macian, P.; Álvarez-Marimon, E.; Sepulcre, F.; Vázquez-Ibar, J.L.; Peralvarez-Marin, A. The Membrane Proximal Domain of TRPV1 and TRPV2 Channels Mediates Protein–Protein Interactions and Lipid Binding In Vitro. *Int. J. Mol. Sci.* **2019**, *20*, 682. [CrossRef]

38. Prescott, E.D.; Julius, D. A modular PIP2 binding site as a determinant of capsaicin receptor sensitivity. *Science (80-.).* **2003**, *300*, 1284–1288. [CrossRef]

39. Mercado, J.; Gordon-Shaag, A.; Zagotta, W.N.; Gordon, S.E. Ca2+-dependent desensitization of TRPV2 channels is mediated by hydrolysis of phosphatidylinositol 4,5-bisphosphate. *J. Neurosci.* **2010**, *30*, 13338–13347. [CrossRef]

40. Doerner, J.F.; Hatt, H.; Ramsey, I.S. Voltage- and temperature-dependent activation of TRPV3 channels is potentiated by receptor-mediated PI(4,5)P2 hydrolysis. *J. Gen. Physiol.* **2011**, *137*, 271–288. [CrossRef]

41. Garcia-Elias, A.; Mrkonjić, S.; Pardo-Pastor, C.; Inada, H.; Hellmich, U.A.; Rubio-Moscardó, F.; Plata, C.; Gaudet, R.; Vicente, R.; Valverde, M.A. Phosphatidylinositol-4,5-biphosphate-dependent rearrangement of TRPV4 cytosolic tails enables channel activation by physiological stimuli. *Proc. Natl. Acad. Sci.* **2013**, *110*, 9553–9558. [CrossRef] [PubMed]

42. Nieto-Posadas, A.; Picazo-juárez, G.; Llorente, I.; Jara-oseguera, A.; Morales-Lázaro, S.; Escalante-Alcalde, D.; Islas, L.D.; Rosenbaum, T. Lysophosphatidic acid directly activates TRPV1 through a C-terminal binding site. *Nat. Chem. Biol.* **2012**, *8*, 78–85. [CrossRef] [PubMed]

43. Snider, J.; Kittanakom, S.; Damjanovic, D.; Curak, J.; Wong, V.; Stagljar, I. Detecting interactions with membrane proteins using a membrane two-hybrid assay in yeast. *Nat. Protoc.* **2010**, *5*, 1281–1293. [CrossRef] [PubMed]

44. Chun, J.N.; Lim, J.M.; Kang, Y.; Kim, E.H.; Shin, Y.-C.; Kim, H.-G.; Jang, D.; Kwon, D.; Shin, S.-Y.; So, I.; et al. A network perspective on unraveling the role of TRP channels in biology and disease. *Pflügers Arch. - Eur. J. Physiol.* **2014**, *466*, 173–182. [CrossRef]

45. Krauß, M.; Haucke, V. Phosphoinositides: Regulators of membrane traffic and protein function. *FEBS Lett.* **2007**, *581*, 2105–2111. [CrossRef]

46. Ferrandiz-Huertas, C.; Mathivanan, S.; Wolf, C.J.; Devesa, I.; Ferrer-Montiel, A. Trafficking of ThermoTRP Channels. *Membranes* **2014**, *4*, 525–564. [CrossRef]

47. Stokes, A.J.; Shimoda, L.M.N.; Koblan-Huberson, M.; Adra, C.N.; Turner, H. A TRPV2-PKA signaling module for transduction of physical stimuli in mast cells. *J. Exp. Med.* **2004**, *200*, 137–147. [CrossRef]

48. Cohen, M.R.; Johnson, W.M.; Pilat, J.M.; Kiselar, J.; DeFrancesco-Lisowitz, A.; Zigmond, R.E.; Moiseenkova-Bell, V.Y. Nerve Growth Factor Regulates Transient Receptor Potential Vanilloid 2 via Extracellular Signal-Regulated Kinase Signaling To Enhance Neurite Outgrowth in Developing Neurons. *Mol. Cell. Biol.* **2015**, *35*, 4238–4252. [CrossRef]

49. Xu, J.; Gu, W.; Ji, K.; Xu, Z.; Zhu, H.; Zheng, W. Sequence analysis and structure prediction of ABHD16A and the roles of the ABHD family members in human disease. *Open Biol.* **2018**, *8*, 180017. [CrossRef]

50. Yokoi, N.; Fukata, Y.; Sekiya, A.; Murakami, T.; Kobayashi, K.; Fukata, M. Identification of PSD-95 Depalmitoylating Enzymes. *J. Neurosci.* **2016**, *36*, 6431–6444. [CrossRef]

51. Herring, B.E.; Nicoll, R.A. Long-Term Potentiation: From CaMKII to AMPA Receptor Trafficking. *Annu. Rev. Physiol.* **2016**, *78*, 351–365. [CrossRef] [PubMed]

52. Kristensen, A.S.; Jenkins, M.A.; Banke, T.G.; Schousboe, A.; Makino, Y.; Johnson, R.C.; Huganir, R.; Traynelis, S.F. Mechanism of Ca2+/calmodulin-dependent kinase II regulation of AMPA receptor gating. *Nat. Neurosci.* **2011**, *14*, 727–735. [CrossRef] [PubMed]

53. Yan, J.-Z.; Xu, Z.; Ren, S.-Q.; Hu, B.; Yao, W.; Wang, S.-H.; Liu, S.-Y.; Lu, W. Protein kinase C promotes N-methyl-D-aspartate (NMDA) receptor trafficking by indirectly triggering calcium/calmodulin-dependent protein kinase II (CaMKII) autophosphorylation. *J. Biol. Chem.* **2011**, *286*, 25187–25200. [CrossRef] [PubMed]

54. Liu, Y.; Cheng, K.; Gong, K.; Fu, A.K.Y.; Ip, N.Y. Pctaire1 phosphorylates N-ethylmaleimide-sensitive fusion protein: implications in the regulation of its hexamerization and exocytosis. *J. Biol. Chem.* **2006**, *281*, 9852–9858. [CrossRef]

55. Palmer, K.J.; Konkel, J.E.; Stephens, D.J. PCTAIRE protein kinases interact directly with the COPII complex and modulate secretory cargo transport. *J. Cell Sci.* **2005**, *118*, 3839–3847. [CrossRef]

56. Dixon-Clarke, S.E.; Shehata, S.N.; Krojer, T.; Sharpe, T.D.; von Delft, F.; Sakamoto, K.; Bullock, A.N. Structure and inhibitor specificity of the PCTAIRE-family kinase CDK16. *Biochem. J.* **2017**, *474*, 699–713. [CrossRef]

57. Rescher, U.; Gerke, V. Annexins - Unique membrane binding proteins with diverse functions. *J. Cell Sci.* **2004**, *117*, 2631–2639. [CrossRef]

58. Gerke, V.; Creutz, C.E.; Moss, S.E. Annexins: linking Ca2+ signalling to membrane dynamics. *Nat. Rev. Mol. Cell Biol.* **2005**, *6*, 449–461. [CrossRef]

59. Dathe, C.; Daigeler, A.L.; Seifert, W.; Jankowski, V.; Mrowka, R.; Kalis, R.; Wanker, E.; Mutig, K.; Bachmann, S.; Paliege, A. Annexin A2 mediates apical trafficking of renal Na+-K +-2Cl- Cotransporter. *J. Biol. Chem.* **2014**, *289*, 9983–9997. [CrossRef]

60. Taylor, J.R.; Skeate, J.G.; Martin Kast, W. Annexin A2 in virus infection. *Front. Microbiol.* **2018**, *9*, 1–8. [CrossRef]

61. Kaczmarek, B.; Verbavatz, J.M.; Jackson, C.L. GBF1 and Arf1 function in vesicular trafficking, lipid homoeostasis and organelle dynamics. *Biol. Cell* **2017**, *109*, 391–399. [CrossRef] [PubMed]

62. Yorimitsu, T.; Sato, K.; Takeuchi, M. Molecular mechanisms of Sar/Arf GTPases in vesicular trafficking in yeast and plants. *Front. Plant Sci.* **2014**, *5*, 1–12. [CrossRef] [PubMed]

63. Price, H.P.; Stark, M.; Smith, D.F. *Trypanosoma brucei* ARF1 Plays a Central Role in Endocytosis and Golgi–Lysosome Trafficking. *Mol. Biol. Cell* **2007**, *18*, 864–873. [CrossRef] [PubMed]

64. Liu, Y.; Kahn, R.A.; Prestegard, J.H. Dynamic structure of membrane-anchored Arf*GTP. *Nat. Struct. Mol. Biol.* **2010**, *17*, 876–881. [CrossRef]

65. Bulley, S.J.; Clarke, J.H.; Droubi, A.; Giudici, M.-L.; Irvine, R.F. Exploring phosphatidylinositol 5-phosphate 4-kinase function. *Adv. Biol. Regul.* **2015**, *57*, 193–202. [CrossRef]

66. Mahajan, K.; Mahajan, N.P. PI3K-independent AKT activation in cancers: a treasure trove for novel therapeutics. *J. Cell. Physiol.* **2012**, *227*, 3178–3184. [CrossRef]

67. Del Bel, L.M.; Brill, J.A. Sac1, a lipid phosphatase at the interface of vesicular and nonvesicular transport. *Traffic* **2018**, *19*, 301–318. [CrossRef]

68. Blagoveshchenskaya, A.; Cheong, F.Y.; Rohde, H.M.; Glover, G.; Knödler, A.; Nicolson, T.; Boehmelt, G.; Mayinger, P. Integration of Golgi trafficking and growth factor signaling by the lipid phosphatase SAC1. *J. Cell Biol.* **2008**, *180*, 803–812. [CrossRef]

69. Hsu, F.; Hu, F.; Mao, Y. Spatiotemporal control of phosphatidylinositol 4-phosphate by Sac2 regulates endocytic recycling. *J. Cell Biol.* **2015**, *209*, 97–110. [CrossRef]

70. Demirsoy, S.; Martin, S.; Motamedi, S.; van Veen, S.; Holemans, T.; Van den Haute, C.; Jordanova, A.; Baekelandt, V.; Vangheluwe, P.; Agostinis, P. ATP13A2/PARK9 regulates endo-/lysosomal cargo sorting and proteostasis through a novel PI(3, 5)P2-mediated scaffolding function. *Hum. Mol. Genet.* **2017**, *26*, 1656–1669. [CrossRef]

71. Usenovic, M.; Knight, A.L.; Ray, A.; Wong, V.; Brown, K.R.; Caldwell, G.A.; Caldwell, K.A.; Stagljar, I.; Krainc, D. Identification of novel ATP13A2 interactors and their role in α-synuclein misfolding and toxicity. *Hum. Mol. Genet.* **2012**, *21*, 3785–3794. [CrossRef] [PubMed]

72. Kehoe, O.; Kalia, N.; King, S.; Eustace, A.; Boyes, C.; Reizes, O.; Williams, A.; Patterson, A.; Middleton, J. Syndecan-3 is selectively pro-inflammatory in the joint and contributes to antigen-induced arthritis in mice. *Arthritis Res. Ther.* **2014**, *16*, 1–14. [CrossRef] [PubMed]

73. Liu, Y.; Echtermeyer, F.; Thilo, F.; Theilmeier, G.; Schmidt, A.; Schülein, R.; Jensen, B.L.; Loddenkemper, C.; Jankowski, V.; Marcussen, N.; et al. The proteoglycan syndecan 4 regulates transient receptor potential canonical 6 channels via RhoA/rho-associated protein kinase signaling. *Arterioscler. Thromb. Vasc. Biol.* **2012**, *32*, 378–385. [CrossRef] [PubMed]

74. Bissen, D.; Foss, F.; Acker-Palmer, A. AMPA receptors and their minions: auxiliary proteins in AMPA receptor trafficking. *Cell Mol. Life Sci.* **2019**, *76*, 2133–2169. [CrossRef] [PubMed]

75. Klaassen, R.V.; Stroeder, J.; Coussen, F.; Hafner, A.-S.; Petersen, J.D.; Renancio, C.; Schmitz, L.J.M.; Normand, E.; Lodder, J.C.; Rotaru, D.C.; et al. Shisa6 traps AMPA receptors at postsynaptic sites and prevents their desensitization during synaptic activity. *Nat. Commun.* **2016**, *7*, 10682. [CrossRef] [PubMed]

76. Breslow, D.K. Sphingolipid homeostasis in the endoplasmic reticulum and beyond. *Cold Spring Harb. Perspect. Biol.* **2013**, *5*, a013326. [CrossRef]

77. Kiefer, K.; Carreras-Sureda, A.; García-López, R.; Rubio-Moscardó, F.; Casas, J.; Fabriàs, G.; Vicente, R. Coordinated regulation of the orosomucoid-like gene family expression controls de novo ceramide synthesis in mammalian cells. *J. Biol. Chem.* **2015**, *290*, 2822–2830. [CrossRef]

78. Kiefer, K.; Casas, J.; García-López, R.; Vicente, R. Ceramide imbalance and impaired TLR4-mediated autophagy in BMDM of an ORMDL3-Overexpressing mouse model. *Int. J. Mol. Sci.* **2019**, *20*. [CrossRef]

79. Wang, J.; Fedoseienko, A.; Chen, B.; Burstein, E.; Jia, D.; Billadeau, D.D. Endosomal receptor trafficking: Retromer and beyond. *Traffic* **2018**, *19*, 578–590. [CrossRef]
80. Abe, K.; Puertollano, R. Role of TRP Channels in the Regulation of the Endosomal Pathway. *Physiology* **2011**, *26*, 14–22. [CrossRef]

Review

Assistance for Folding of Disease-Causing Plasma Membrane Proteins

Karina Juarez-Navarro, Victor M. Ayala-Garcia, Estela Ruiz-Baca, Ivan Meneses-Morales, Jose Luis Rios-Banuelos and Angelica Lopez-Rodriguez *

Facultad de Ciencias Quimicas, Universidad Juarez del Estado de Durango, Durango C.P 34000, Mexico; karinajuarezn@hotmail.com (K.J.-N.); victor.ayala@ujed.mx (V.M.A.-G.); eruiz@ujed.mx (E.R.-B.); ivan.meneses@ujed.mx (I.M.-M.); qfbrios@gmail.com (J.L.R.-B.)
* Correspondence: angelica.lopez@ujed.mx; Tel.: +52-618-827-1340

Received: 9 February 2020; Accepted: 21 April 2020; Published: 7 May 2020

Abstract: An extensive catalog of plasma membrane (PM) protein mutations related to phenotypic diseases is associated with incorrect protein folding and/or localization. These impairments, in addition to dysfunction, frequently promote protein aggregation, which can be detrimental to cells. Here, we review PM protein processing, from protein synthesis in the endoplasmic reticulum to delivery to the PM, stressing the main repercussions of processing failures and their physiological consequences in pathologies, and we summarize the recent proposed therapeutic strategies to rescue misassembled proteins through different types of chaperones and/or small molecule drugs that safeguard protein quality control and regulate proteostasis.

Keywords: proteostasis; quality control; misrouting; chaperones

1. Introduction

Currently, we understand the plasma membrane (PM) not as a simple lipid bilayer protecting the cells or surrounding the cytoplasm but as a collection of stably folded membrane proteins (MPs) in an asymmetric arrangement. In the PM, peptides interact with the lipid bilayer hydrocarbon core, the bilayer interface, and water in a minimum free energy state, forming complex and dynamic protein–lipid structures that participate directly as messengers or regulators of many signal transduction cascades. Regulation of these complex structures is essential for life and health [1–3]. MPs are difficult to study in vitro for many reasons, such as their flexibility, instability, and relatively hydrophobic surface. Since the first MP structure was published in 1985 [4], our knowledge has increased slowly but steadily. However, many aspects of the cell membrane are incompletely understood, including its lipid–protein organization and its stability to allow substances that meet strict criteria to transit unaided or through protein transporters, maintaining the intracellular/extracellular balance of substances according to physiological conditions.

Understanding how PM proteins are regulated from their synthesis to their final localization could provide further insight into the mechanisms of certain cellular events, such as folding, molecular sorting, and intracellular transport, that take place in lipidic membranes after protein synthesis. A wide range of genetic diseases is related to MP misfolding. Because the mutant proteins are either retained/accumulated intracellularly or are dysfunctional at the PM, signaling cascades mediating various physiological processes are mainly affected. In this review, we focus on PM proteins, analyzing their synthesis, folding, and trafficking in addition to the mechanism allowing their expression and stability at the membrane, and highlighting approaches by which natural and synthetic chaperones can be used to rescue misfolded phenotypes as strategic therapies to treat misfolding-related diseases.

2. Membrane Proteins

The external boundary of the cell is the plasma membrane (PM), and organelles are delimited by intracellular membranes. Lipids are essential components of all cell membranes. The composition of lipids and MPs may differ substantially depending on the function, organelle type, cellular location, or tissue level with which they are associated [1–3]. In many cells, phospholipids (glycerophospholipids and sphingolipids) are the cellular building blocks, while non-phospholipids are important regulators of lipid organization [1]. Cholesterol is the most abundant non-phospholipid in mammalian biomembranes. In human brain cells, cholesterol is a major constituent; it is critical for brain development. Cholesterol depletion leads to central nervous system pathologies such as Huntington's [5] and Alzheimer's [6], among other diseases [7,8]. Lipid content of PM influences the ion channel electrostatic environment, leading to an indirect modulation of ionic current and charge movement by modifying the transmembrane. Different studies have experimentally demonstrated that voltage-gated potassium (Kv) channels are sensitive to cholesterol [7–11] and phospholipid content [12–15].

Indeed, lipid distribution is highly regulated and not random across different membranes as phosphoinositides (PIs) show a clear demarcation in the cell. The eight members of the mammalian PI family (PI, PI3P, PI4P, PI5P, PI(4,5)P2, PI(3,4)P2, PI(3,5)P2, and PI(3,4,5)P3) play critical roles in modulating biological processes, such as gene expression, signaling, and membrane and cytoskeletal responses, and in membrane trafficking, among others. Thus, whereas some members, such as PI3P and PI(3,5)P2, along with their effectors and specific phosphatases are mainly present at the early and late endosome, respectively [16], PI(4,5)P2 is essential for endocytosis, exocytosis, and the regulation, adhesion, and assembly of membrane proteins [17,18]. Because of the specific distribution of PI family members in the cell, it is reasonable to fulfill important roles for MP location and function. Recently, multiple roles of lipids in ion channels and transporter regulation have been demonstrated, highlighting the importance of lipid–protein interactions for the cell physiology [19,20].

The abundance of MPs in the cell is low compared to the total protein population; the proportion of putative MPs predicted from sequenced genomes is between 20% and 35% [21,22]. PM proteins carry out diverse functions such as transporting nutrients to cells, regulating the exchange of bioactive molecules and receiving chemical signals from the extracellular space, activating signaling pathways in response to different stimuli, allowing the translation of chemical signals into intracellular activity, enhancing intercellular interactions, and sometimes promoting cell anchoring in a particular location [23–25].

MPs are grouped into two broad categories based on the nature of their interactions: (1) Integral MPs, also called intrinsic proteins, are integrated into the membrane; many of them can span the entire lipid and include more than one linked transmembrane domain fully embedded in lipid bilayers (also called transmembrane proteins). Examples of integral MPs include aquaporins, ion channels, transporters, and pumps. (2) Peripheral MPs, or extrinsic proteins, are entirely outside the membrane, indirectly bound to it by weak molecular interactions (e.g., ionic, hydrogen, and/or Van der Waals bonds), with integral MPs or the polar head groups on lipids. These MPs include phospholipase C, alpha/beta hydrolase fold, annexins, and synapsin I [26–28].

The mechanisms governing the selection and localization of PM proteins are strongly controlled; MPs have additional sequences (e.g., stop–transfer sequence membrane-spanning regions, glycosylphosphatidylinositol (GPI) anchors) that allow their integration into endoplasmic reticulum (ER) membranes. As reticular membranes move to the Golgi apparatus and finally to the PM, transmembrane proteins, such as ion channels and transporters, remain integrated with the internal membranes and then associate with the external membrane or stay partially embedded in one leaflet of the bilayer [29–32]. Caveolins (Cavs) are a good example of integral proteins partially embedded in one leaflet of the bilayer. The amino and carboxy terminal domains of Cavs flank a central hydrophobic region; hence, the formation of a hairpin in the lipidic bilayer is suggested by the aminoacidic sequence of this protein. Cavs are essential proteins for the formation of PM invaginations called Caveolae ("little caves"). The co-translational ER membrane insertion of Cavs depends on the signal recognition sequence [33,34]. Cavs are exported in a vesicle-dependent manner from the ER to Golgi. After post-translational modifications, caveolae transport oligomeric Cavs embedded into cholesterol-rich membranes. At the cell surface, Cavs interact with protein adaptors (Cavins) to assist membrane curvature. The lipidome of caveolae depends on the cell type and physiological condition. Lipid–protein interaction in caveolae may influence protein conformation, protein interactions, or ligand affinity, affecting cascaded signal transduction and cell physiology [35].

Protein Biosynthesis

When mRNA reaches the cytosol, two ribosomal subunits associate with the initiation codon through eukaryotic initiation factors (eIFs) to integrate the translation initiation complex [36]. After the signal peptide denoting an MP is detected by the protein/RNA complex—called signal recognition particle (SRP)—the new protein will be inserted into the ER membrane, mediated in a co-translational or post-translational manner [37]. In general, all MPs are assembled in the ER, achieving their tertiary and quaternary structures.

In the co-translational pathway, many MPs and secretory proteins are formerly expressed as a pre-protein with an N-terminal topogenic sequence, which is a crucial signal peptide for protein localization and for protein insertion and orientation in cellular membranes. Commonly, 15 to 30 amino acids conform to the signal peptide; however, length can be up to 50 residues. This peptide sequence is typically cleaved off co-translationally [38]. Table 1 displays several signal sequences reported for proteins located in the ER and PM.

Table 1. Signal peptide for protein localization in the endoplasmic reticulum and plasma membrane.

Sequence	Protein	Location	Organism	Reference
VVQAITFIFKSLGLKCVQFLPQVM PTFLNVIRVCDGAIRE.	mTOR.	Endoplasmic reticulum.	Homo sapiens; Mus musculus; Rattus norvegicus.	[39]
HALSYWKPFLVNMCVATVLTAGA YLCYRFLFNSNT.	PTP-1B.	Endoplasmic reticulum.	Homo sapiens.	[40]
MEAMWLLCVALAVLAWG.	GlcNAc-PI.	Endoplasmic reticulum.	Homo sapiens.	[41]
IPHDLCHNGEKSKKPSKIKSL FKKKSK.	STIM2.	Endoplasmic reticulum.	Homo sapiens; Mus musculus.	[42]
GVMLGSIFCALITMLGHI.	Cosmc.	Endoplasmic reticulum.	Bos taurus; Homo sapiens; Mus musculus; Rattus norvegicus.	[43]
MRLLLALLGVLLSVPGPPVLS.	FGFR4.	Plasma membrane.	Homo sapiens.	[44]
MDCRKMARFSYSVIWIMAIS KVFELGLVAG.	TDGF.	Plasma membrane.	Homo sapiens.	[45]
MPAWGALFLLWATAEA.	(GP)IX.	Plasma membrane.	Homo sapiens.	[46]
LRCLACSCFRTPVWPR.	prRDH.	Plasma membrane.	Bos taurus.	[47]
MGCGCSSHPE.	Lck.	Plasma membrane.	Homo sapiens; Aotus nancymaae.	[48,49]
VTNGSTYILVPLSH.	FSHR.	Plasma membrane.	Homo sapiens.	[50]
AETENFV.	M3 mAChR.	Plasma membrane.	Homo sapiens; Gorilla gorilla gorilla; Pan troglodytes; Pongo pygmaeus.	[51,52]

In general, even when sequence conservation among signal peptides is low, they have common secondary structural features, including a hydrophilic region at the N-terminal—which is frequently positively charged, a central hydrophobic domain, and a site for signal peptidase cleavage located in the C-terminal region; signal peptides can also display extended N-regions or hydrophobic regions [38,53].

After the signal peptide denoting an MP is detected by the SRP, it interacts with the SRP receptor in a GTP-dependent manner and undergoes several structural modifications. Then the SRP–ribosome complex docks to the translocon—a channel composed of proteins crossing the lipid bilayer to the ER lumen [10,17–21]—and translocation is resumed. the translocon is closed and the ribosome dissociates. The nascent protein is unloaded into the translocation channel (or translocon), GTP hydrolysis occurs, and the SRP receptor is free for another cycle [41]. Along with peptide translocation, signal peptide cleavage is mediated by the signal peptidase complex. Cleavage frequently occurs in a co-translational way [42–44], but it can also happen at some point between the early co-translational and late post-translational stages [45]. When translocation finishes, the translocon is closed and the ribosome dissociates.

When the ER protein in the lumen is translocated across the membrane, the protein is shifted laterally for anchoring within the phospholipid bilayer, allowing the protein to be integrated or assembled with other proteins into the ER membrane [54,55]. During translocation, enzymes (e.g., signal peptidases and oligosaccharyltransferase) can associate with the protein to cleave the signal peptide or N-glycosylate the translocating nascent chain [56,57].

In a conventional protein trafficking pathway after ER processing, vesicles favor the anterograde protein traffic toward the Golgi apparatus to further fuse transport vesicles with the target membrane, allowing protein insertion into the PM. Vesicular trafficking also occurs in an anterograde way to support PM protein quality control and cell integrity. A growing number of proteins have been associated with an unconventional protein trafficking pathway where proteins are delivered to the surface of the PM in an ER- or Golgi-independent manner. Exceptionally, most of the proteins following the unconventional trafficking pathway lack the classical N-terminal signal peptide, although if the signal peptide is recognized, proteins bypass the Golgi to reach the PM or be excreted [58].

Particularly, the C-tail anchored MPs—called tail-anchored (TA) proteins—constitute 3%–5% of the eukaryotic membrane proteome [37,59]; they lack the classical N-terminal signal for membrane insertion because they constitute a single stretch of hydrophobic amino acids where the membrane-interacting region is near the COOH terminus. After translation ends, a hydrophobic region is recognized and captured by specific cytosolic chaperones forming a pre-targeting complex that drives the opening of an ER membrane receptor, allowing TA protein insertion into the ER by a GTPase-dependent step [37,60,61]. TA proteins are found essentially in the intracellular membranes, carrying out various enzymatic and regulatory functions in the cellular metabolism, including apoptosis and protein quality control processes, besides protein localization and membrane traffic [62].

3. Membrane Protein Folding

In addition to the folding guided through the amino acid sequence, cytosolic regions may interact with intracellular proteins or chaperones for proper folding. Molecular chaperones are proteins that interact with a nascent protein to assist the stabilization of the native conformation, allowing the protein to remain in the intermediate states for longer during the folding process, but chaperones are frequently absent in the final functional structure [63]. The major ER-resident chaperones are proteins of the heat shock protein (Hsp) family [64], lectin chaperones, calnexin, and calreticulin [65].

Several Hsps are usually located in the ER; however, different types and levels of chaperone genes can be expressed under stress conditions or indifferent cellular stages [65–70].

In addition, chaperones assist the translocation machinery and play a role in the retrograde transport of aberrant proteins destined for proteasomal degradation [70,71], regulating the unfolded protein response (UPR) [72].

Along with an abundance of cytosolic and ER-specific molecular chaperones, the ER contains several classes of folding enzymes. *Prolyl cis-trans* isomerases, which via *cis-trans* isomerization of proline residues constrain the flexibility of the peptide backbone favoring the formation of disulfide bonds, catalyze co-translational and post-translational modifications important for protein folding [73–75]. Different kinds of chaperones are dissociated or associated with proteins throughout the protein maturation process [76,77].

Although an entire machinery of molecules participates in the protein folding process, misfolding sometimes happens despite the cell efforts to prevent it. Mutations and protein translation errors are frequently associated with protein misfolding. Nevertheless, different factors (e.g., age-related errors, exposure to environmental stress conditions, or lack of chaperone availability) can also induce aberrant folding [78]. To avoid misfolded protein aggregates, cells have developed sophisticated quality control mechanisms preventing dysfunctional proteins from reaching their destination. Control quality mechanisms are mediated by chaperone-dependent disaggregation and refolding systems and/or systems regulated through selective proteolysis. However, when a system fails, misfolded toxic aggregates lead to severe human diseases, such as neurodegenerative diseases (e.g., Alzheimer's, Parkinson's, Creutzfeldt–Jakob, and Huntington's), diabetes, and cancer, among others [79–81].

The structural integrity of proteins must be constantly monitored by quality control mechanisms throughout their life—from translation in the ribosome to their arrival at the functional location in the cell. Once the proteins reach their final destination, a subsequent monitoring system ensures protein integrity; if at some point in their life cycle, proteins are recognized as terminally misfolded, they will be eliminated.

Proteins are synthesized on cytosolic ribosomes, but the ER is the main entry gate for secretory proteins expressed in intracellular organelles (including the ER), the PM, and cellular exterior. Hence, the ER features the first-line cellular quality control system (QCS) as it must ensure proper folding and assembly for proteins [63,82,83].

3.1. Quality Control Systems for Membrane Protein Folding

Cells have different QCSs to remove unfolded proteins. The UPR is a fundamental signaling pathway to keep cell homeostasis; through this pathway, unfolded proteins are exported from the ER and degraded in lysosomes, therefore increasing the ER folding capacity. When misfolded MPs are retained in the ER, the latter becomes stressed, and the UPR pathway is activated. The UPR comprises multiple strategies acting in parallel and/or in series to restore normal ER functioning [84].

In response to ER stress, major branches of the UPR are activated with the following aims: (1) increase the biosynthetic capacity—upregulate the expression of ER-resident chaperones—to prevent protein aggregation and facilitate correct protein folding; (2) control transcription, regulate mRNA abundance by stimulating or inhibiting transcription or by enhancing or compromising mRNA stability; (3) decrease the biosynthetic burden (attenuate translation) to reduce the transit of proteins through the ER, while the synthesis of membrane lipids increases the ER volume; (4) translocate misfolded proteins out of the ER; (5) remove misfolded proteins within the ER (by lysosomal/proteasomal degradation or ER autophagy) [85,86].

These branches include at least three mechanistically different components of the UPR: the RNA-dependent protein kinase-like ER kinase (PERK), activating transcription factor 6 (ATF6), and inositol-requiring ER-to-nucleus signal kinase 1 (IRE1). Coordinated actions of these proteins modulate gene expression, affecting the synthetic and secretory pathways, cell fate, and the metabolism of proteins, amino acids and lipids by activating specific transcription factors (e.g., ATF4, ATF6N, and X- box-binding protein 1 [XBP1], respectively) to lower ER stress [87]. A wide range of dysfunctions would otherwise be lethal if not for this intervention. When it becomes clear that a misfolded protein cannot be properly refolded, cellular stress persists, and the UPR negatively impacts the health of the cell and induces apoptosis [88–90]. As part of the UPR response, the transmembrane protein kinase PERK inhibits the translation of new proteins. After sensing ER stress, oligomerization of the luminal domain (N-terminal) of PERK facilitates autophosphorylation. After PERK is processed, it phosphorylates the α subunit of eukaryotic initiation factor 2 (eIF2α), which induces a transient attenuation of protein translation along with the activation of stress-responsive transcription factors to stimulate the expression of chaperones, oxidative response genes and autophagy/apoptosis genes, among other UPR-related proteins [53–55,91,92].

IRE-1 represents the most conserved signaling pathway. Through this pathway, chaperone expression is increased to modulate the ER-associated degradation (ERAD) pathway, responsible for a common process that clears the ER from potentially harmful species. ERAD machinery drives the retrotranslocation of misfolded proteins to the cytosol, where ubiquitin/proteasome proteolysis occurs [78,89]. ATF6 and IRE-1 interaction regulate the quantitative and qualitative expression of XBP1 to compensate for unfolded protein accumulation. After the UPR response is activated, ATF6 is transported to Golgi, where it is further enzymatically cleaved. The released fragment acts as a transcription factor that regulates the expression of proteins such as IRE1 and BiP chaperone [93,94].

Unfolded proteins stuck in the ER are eventually degraded; therefore, they must be retranslocated into the cytosol. Proteins require a signal, like ubiquitination, to be recognized for degradation, thus ensuring their delivery to the proteasome. However, a peptide signal for degradation is unclear. Experimental evidence shows that some molecular chaperones or protein disulfide isomerase homologs can associate with degradation substrates to prevent aggregation and escort selected polypeptides from the ER to the proteasome or lysosomes [95]. After retrotranslocation, several ubiquitin-binding proteins guide degradation substrates to the proteasome [96]. Even when glycosylation and protein degradation are evidently associated, it is less clear how non-glycosylated proteins are recognized for degradation. As the efficiency of binding through the calnexin/calreticulin cycle is dependent upon the oligosaccharide structure, proteins that are not properly glycosylated may activate a degradation pathway known as enhancing α-mannosidase-like protein (EDEM) [97]. Since the membrane protein degradation mechanism is not yet understood, it may start at the protein soluble parts after dislocation from the retrotranslocon or through a direct protein excision from the lipidic membrane. Analysis of MP degradation fate demonstrated that undegraded molecules accumulate in the cytoplasm when proteasome function is compromised, suggesting that transmembrane segments might be solubilized from the ER membrane before proteasome-mediated degradation [98].

Along with protein degradation mediated by proteasomes or lysosomes, ER autophagy (ER-phagy) can occur as an alternative to de-stress ER, removing aberrant portions of the ER containing abnormal proteins [99,100]. ER stress-mediated autophagy is a cellular catabolic process in which misfolded proteins, protein aggregates, and damaged ER regions are transported to the lysosome for degradation. Lysosomal degradation is induced by a set of hydrolases working in an acidic environment; afterwards, building blocks are recycled by the cell. The mechanisms by which MPs are tagged for lysosomal degradation are known when the translocation of cytosolic proteins is associated with a specific degradation signal (the KFERQ sequence motif) [101].

Evidently, the UPR response affects not only ER-related proteins but also genes associated with different cellular processes, including metabolism and inflammation; thus, gene regulation cannot always be explained through a canonical mechanism that only considers ER factors directly affecting protein processing [102,103]. UPR transcriptional output can be supported by parallel non-canonical stress-sensing mechanisms that may modulate the canonical mechanism. The translation of proteins that are not related to ER function may be implicated, as well as the self-association of proteins creating stress-specific scaffolds integrated by a multimeric protein complex. ER stress can also induce transcriptional cascades, expanding the expression of UPR gene regulatory networks and regulatory complexes not expressed constitutively. For example, the association of PERK and IRE-1 may create stress-specific scaffolds, inducing splicing mechanisms on XBP1 protein expression and affecting the transcriptional regulation [104]. IRE-1 can also recruit the TNF receptor-associated factor 2 (TRAF2), inducing activation kinases (e.g., IKK, JNK, ASK1, p38 MAPK, and ERK) that are important to determine the cell's fate (survival/apoptosis) [105]. Another emerging point of regulation is related to microRNAs, the short (~22 nt), single-stranded RNAs binding to complementary mRNAs to inhibit protein translation [106]. In general, the expansion of UPR signals impact cell fate.

To prevent issues related to the quality control of protein folding, the cell developed QCSs beyond ER checkpoints; for instance, in the Golgi apparatus, Rer1 and ERp44 proteins recognize immature proteins and facilitate their retrograde trafficking to the ER [107,108]. After passing the Golgi QCS, MPs embedded into transport vesicles can continue towards their functional locations. PMs also seem to have their own QCSs, including endocytic adaptors and ubiquitination systems, to monitor the MP structural integrity [109].

Therefore, intracellular trafficking is highly regulated, even when proteins may have particular motifs for exportation. If a QCS detects structural frustration indicating misfolding [110], proteins will be stuck in traffic. Misfolded proteins can undergo a refolding process, although proteins unable to fold may induce ER or mitochondrial stress, activating the degradation pathway to release the stress. After passing the QCS at the ER and Golgi, protein composition at the PM represents a complicated balance of membrane delivery, endocytosis, and recycling mechanisms [82] (Figure 1).

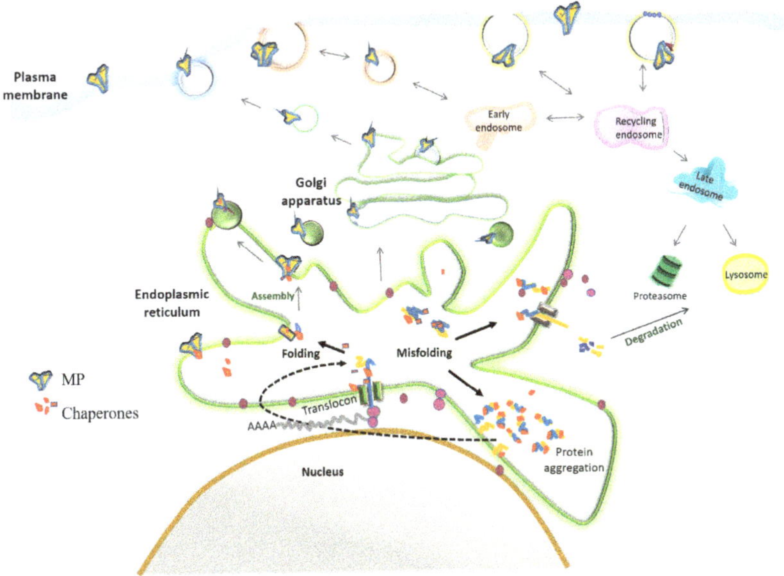

Figure 1. Trafficking of membrane proteins. As soon as the new membrane protein (MP) starts translocating to the ER, it will associate with chaperones and other proteins, assisting in getting a structural conformation into the lipid membranes. Well-folded proteins will continue trafficking through the endomembrane system to and from the PM, while misfolded proteins are re-directed to ER protein folding or degradation pathways, reducing their secretion to the extracellular space where they could further misfold or aggregate into proteotoxic conformations.

Long- and short-distance communication can take multiple vesicular forms, generally created through a membrane budding and pinching off mechanism. For MPs, vesicular transport is a milestone to safeguard proper folding and integrity while trafficking to the final destination. After ER processing, proteins are transported by coated vesicles (60–90 nm in diameter). The coat is a protein complex integrated by four subunits: Sec23/24-Sar1 selects cargo while Sec13/31 deforms the membrane to induce the budding. Some of the proteins incorporated into the vesicles can act as receptors for soluble proteins, thus favoring their packaging. After the vesicle buds from ER, translocation to the Golgi apparatus occurs, allowing post-translational modifications [111–113].

All cells contain a subset of membranous vesicular/tubular carriers formed by a direct budding of membranes. Vesicles are responsible for several cellular processes involving intracellular trafficking, including endocytosis and exocytosis. Primary endocytic vesicles can fuse with early endosomes to continue toward the protein maturation process—via a constitutive recycling pathway—or to prepare them for transportation to lysosomes [114]. Vesicles can be classified into three groups based on size and biogenesis: Apoptotic bodies released as an apoptotic response (800–5000 nm in diameter), ectosomes released directly from the PM (100–1000 nm in diameter), and exosomes originated from the inward budding of endosomes into multivesicular bodies (late endosomes; 30–100 nm in diameter) [115]. Despite the vesicular origin and size, there is still no reliable way to distinguish between ectosomes and exosomes, and the function may be quite analog. Particularly, some of the exosomes carry cargo molecules directly into the lysosomal compartment for degradation. Exosomes also form intraluminal vesicles to transport molecules (e.g., proteins, RNA [116,117] or Hsp chaperones [118]), which along with the cargo of extracellular vesicles may affect organelle function or modulate recipient cell function, thus contributing to the molecular intercellular transmission [119]. The recognition and packaging of cargo proteins result from multiple cooperative interactions between accessory proteins and lipids. Distinct ER-to-Golgi forward-trafficking signals have been identified on cargo or cargo receptor proteins. In potassium channels (Kir 1.1 and Kir2.1), the export signaling motif is not required for channel folding, assembly, or gating, but it is crucial for ER export [120].

3.2. Membrane Protein Modifications

After the new proteins are correctly folded and assembled in the ER, they travel towards the Golgi apparatus into transport vesicles, where they usually undergo different types of post-translational processing; however, protein modifications occurring co-translationally in the ER or post-transductionally in the Golgi apparatus also play an integral and crucial role in protein trafficking and function [112].

Some protein modifications, such as glycosylation, can reduce protein dynamics by increasing stability and favoring protein trafficking. N-glycosylation is the most common process to add sugar moieties to proteins. Glycosylation often starts while MP is translocated into the ER, and a preassembled polymannose oligosaccharide is transferred to the luminal N-aminoacidic residue of the classic motif including asparagine-X-serine/threonine(N-X-S/T) as a consensus sequence. Once in the Golgi, some enzymatic reactions can add sugar moieties or just modify the preexisting glycan tree complexity [121–123].

Phosphorylation is the enzymatic transference of phosphate from the ATP molecule on the side chains of serine, threonine, or tyrosine residues. Phosphorylation can occur co- or post-translationally in a reversible process affecting a wide variety of processes, including protein trafficking, clustering, conformation, and protein–protein interactions [124]. There are a few well-studied phosphorylation-induced trafficking examples. For instance, serotonin transporter (SERT) trafficking can be regulated when serine and threonine phosphorylation favors the recruitment of membrane skeleton adaptor protein Hic-5, inducing the actin-dependent endocytosis [125–127]. Exceptionally, phosphorylation at tyrosine residue in Kv1.3 triggers opposite effects depending on the protein life-stage, inducing either protein surface targeting or endocytosis of the channel [128].

The formation of reactive oxygen and nitrogen species (ROS/RNS) normally induces the post-translational oxidative modification of proteins, usually affecting cysteine residues due to their highly reactive thiol group. ROS/RNS can affect the thiol group of cysteine residues. In the mammalian proteome, two amino acids contain sulfur residues: cysteine and methionine. Particularly, the thiol group of cysteine enables multiple oxidation states allowing redox modifications that contribute to the signaling cascade specificity [129]. Many types of cysteine oxidative modifications can be reversed depending on the physiological condition of the cell [130]; protein modification through cysteine residues senses and transduces signaling cascades and regulates biological outcomes. Post-translational modifications like phosphorylation, glycosylation, and ubiquitination can combine with redox regulation to control cellular physiology. Even though many redox mechanisms affecting PM protein trafficking are still unknown, these mechanisms are interesting and promising, especially to understand the progression of many pathologies such as cancer and cardiovascular diseases [131].

At least six types of lipids, including fatty acids, sterols, isoprenoids, GPI anchors, and lipid-derived electrophiles, can attach to the cysteine, serine, or lysine residues of proteins. The covalent but reversible attachment of fatty acid with cysteines via a thioester linkage is called S-acylation, whereas the covalent interaction between lipids and proteins is called protein lipidation. This process occurs co- or post-translationally, and its deregulation has been linked to different diseases, including metabolic diseases, neurological disorders, and cancers [132]. Reversibility of protein lipidation allows multiple regulatory scenarios during protein lifetime. In most cases, the acyl chain attached to the protein is unknown; however, there is some experimental evidence showing that palmitoylation or myristoylation can modulate the trafficking of some potassium channels [133–135]. So far, covalently bound proteins-cholesterols are uncommon except in Hedgehog proteins, Smoothened, and Hh pathway co-receptor [132]. Protein phospholipid modification is also rare; to date, the only known example is the autophagy-related protein Atg8/LC3 [136,137].

Ubiquitination is a physiologically common interaction among proteins that describes the covalent attachment of ubiquitin to Lys residue in a target protein. Ubiquitination complexity is related to degradation, particularly in the ubiquitin–proteasome and autophagy–lysosome pathways. Alterations in the ubiquitin system lead to the development of many diseases [138].

3.3. Membrane Protein Expression and Stability

In general, protein misfolding can generate two different protein phenotypes (Figure 2): (1) Mislocalized proteins, which commonly induce cellular stress due to improper degradation [139] or structural alterations that establish novel toxic functions, sometimes inducing ER and mitochondrial stress leading to apoptosis [139–141] or amyloid accumulation [142–149], or (2) dysfunctional proteins due to an increased (overexpressed) or decreased (underexpressed) amount of protein reaching the PM. This phenotype occasionally disrupts protein stability at the PM (rendering proteins unstable) because of increased turnover via endocytic and recycling mechanisms; in addition, misfolded proteins that reach the PM frequently exhibit altered (atypical) functionality, such as underexpression, overexpression, or complete loss of function [150,151]. In some cases, the functionality of misfolded proteins is lost because of their mislocalization rather than the loss of the intrinsic ability of the mutant receptor to interact with its ligands or effectors.

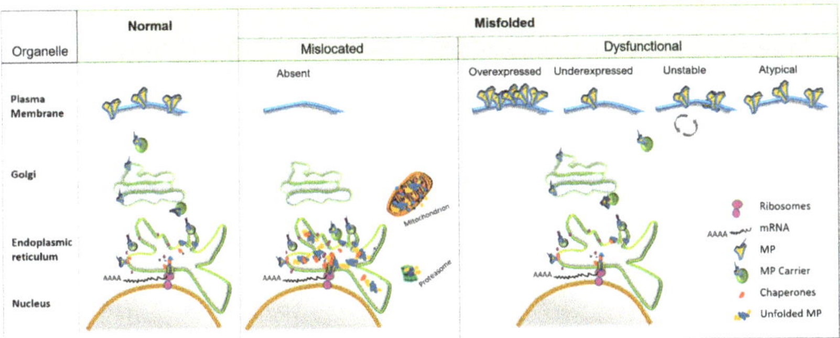

Figure 2. Misfolded membrane protein phenotypes. Simplified scheme of membrane proteins (MP) with folding problems. Normal: As soon as a nascent protein starts being translated at the ribosomes and translocated to the ER, chaperones will assist in folding, and well-folded proteins will travel in lipidic vesicles (MP carrier) from the ER to the Golgi apparatus to reach the plasma membrane, where they undergo a constant recycling cycle. Mislocated: Abnormally folded proteins will selectively be excluded from transport vesicles for accumulation in the lumen of the ER, triggering a heightened state of ER stress; to relax the stressed ER, the proteasomal degradation of proteins will be induced. Protein aggregates can also induce mitochondrial stress, promoting apoptosis. Dysfunctional: Function of misfolded protein reaching the PM can be affected due to protein over- or underexpression; if protein expression seems normal, protein stability in the plasma membrane may be affected, increasing the recycling turnover rate, or function can be atypical (i.e., non-functional or having a differently modulated function).

The native forms of most proteins are in a maximum stable thermodynamic state [152]. However, many proteins are metastable, meaning that thermodynamic stability can also be achieved by using alternative folding pathways, thus inducing the formation of inactive conformations; this mechanism is crucial for regulating the biological function of proteins [152–157]. Therefore, under mild destabilizing conditions, proteins have an inherent tendency to misfold and aggregate and, hence, lose functionality.

Protein levels are thus tightly regulated intracellularly and extracellularly; however, under some circumstances, such as aging, mutation, and environmental stress [158–160], misfolded proteins can have inappropriately exposed hydrophobic surfaces that are normally buried in the interior of the protein, leading to nonnative conformations that can interact with each other to form aggregates [161].

4. Physiological Consequences of Protein Misfolding

4.1. Effects on Membranes

The arrangement of amino acids and the exposure of hydrophobic residues are important factors affecting the ability of a protein to interact with a lipid membrane and can induce membrane fusion and destabilization. Several studies have shown that specific lipids (also called "lipochaperones") can perform a chaperone-like function in insertion and folding, guiding the assembly of MPs [161,162]. Changes in the cell membrane affected by misfolded proteins can vary depending on the nature of the proteins and the type of lipids involved. Some proteins, particularly those characterized by electrostatic interactions, merely disrupt the membrane by the binding of positively charged amino acid residues to negative or polar lipid head groups. Such disturbances are likely reversible and short-term for some proteins [163–165]. In addition to electrostatic disturbance, the insertion of misfolded protein can affect membranes in several ways:

(A) Vesicle formation. The exit signals that direct proteins out of the ER for transport to the Golgi apparatus and beyond are poorly understood. Signal-dependent transport is mediated by the recognition of discrete export signals on the cargo molecule by specific receptors concentrated at specific vesicle binding sites. Thus, to be exported from the ER to the functional location, proteins must be properly folded and completely assembled, while misfolded proteins will remain in the ER bound to chaperone proteins, which may obscure the exit signals or anchor the proteins to the ER, disturbing MP trafficking into vesicles [166–169].

(B) Loss of membrane integrity. Different lipid compositions can alter the bilayer's natural thickness. To shield hydrophobic surfaces from the aqueous environment, some membrane tension can be induced if the hydrophobic surface of an MP is thicker or thinner than the hydrocarbon core of the bilayer [170,171]. The arrangement of hydrophobic amino acid residues in a protein is fundamental to favor protein–lipid interactions, and misfolding may affect the shape or thickness of the membrane around the protein [172].

(C) Formation of pathologic ion channels. To minimize exposure of the hydrophobic regions to the aqueous medium, a hydrophobic protein may change its structural conformation to act as a bridge, allowing specific lipid interactions in the membranes and thus increasing the likelihood of ion channel formation, regardless of the native protein's secondary structure [173].

4.2. Effect on Protein Structure and Assembly

The composition and spatial arrangement of the subunits integrating functional proteins are a prerequisite for protein function and trafficking to the cell membrane [31,171]. The assembly of diverse proteins to perform a specific function is possible depending upon cell type-specific subunit expression, as subunit assembly can be controlled in a developmentally regulated manner or in response to cellular activity. Protein assembly relies on subunit stability in the ER, intersubunit affinities, and potential subunit diffusion within the ER membrane. The homomeric structure is the simplest type of protein assembly; it can be integrated by self-assembly of repeated copies of the same subunit or formed from heteromeric complexes made of multiple distinct protein subunits [174]. Incompletely assembled complexes are usually selectively retained. The association of two or more polypeptide chains to form nonfunctional structures is defined as misassembly and is usually, by definition, a consequence of misfolding [175].

Misfolding and misassembly typically occur in the ER but can also affect other organelles, for example, (1) defective peroxisomal assembly is associated with inherited human diseases (commonly called peroxisomal biogenesis disorders), such as the severe cerebrohepatorenal Zellweger syndrome (ZS). PBDs are mainly caused by mutations in *PEX* genes codifying for peroxins, proteins responsible for normal peroxisome assembly and functions [176,177]. (2) Some transport vesicles carry cargo molecules transporting misassembled proteins to the Golgi apparatus or beyond. Proteins retained in the Golgi apparatus can also be targeted by lysosomes for degradation, suggesting that additional quality control checkpoints could act in this compartment [178,179]. (3) Misassembled proteins can be correctly targeted for degradation and transported out of the ER. When the degradation machinery fails, misassembled proteins accumulate in the cytosol as aggresomes [180].

Even when the loss of proteostasis underlies aging and neurodegeneration characterized by the accumulation of protein aggregates and mitochondrial dysfunction, the relationships among these negative factors are unclear. Evidence suggests that some cytosolic proteins susceptible to aggregation are imported to the mitochondria, which seems to act as a guardian of cytosolic proteostasis [181].

Conformational changes in the structure of a mutated protein lead to the formation of a partially folded intermediate. Intermediate states in protein folding naturally occur even in wild-type proteins. Folding/unfolding transitions thermodynamically and kinetically follow multiple discrete steps that can sometimes provide a starting point for aggregation [182]. When protein monomer aggregates produce fibrillar structures rich in β-strand conformations, the formed structures are usually called amyloid aggregates. Amyloidogenic proteins have been related to normal physiological functions such as bacterial biofilm formation or regulation of synthesis and storage of melanin in human. Amyloid aggregates are frequently related to neurodegenerative diseases that include Alzheimer's disease, Parkinson's disease, various tauopathies, and other human disorders [183]. Although there is no clear evidence that PM protein can form amyloid aggregates, experimental evidence shows pore formation in the bilayer due to the amyloid aggregate–lipid interaction [184–186], toxically affecting membrane permeabilization and ion homeostasis. In some cases, misfolded proteins and their aggregates can be re-solubilized and re-folded by protein chaperones or degraded by the ubiquitin–proteasome pathway [187]. Scientists are starting to understand the molecular mechanisms underlying the development of misfolding diseases. Mutations affecting the activity of chaperones to prevent protein misfolding, along with the production of harmful proteins and the formation of misfolded protein aggregates, may induce the toxicity associated with these pathological disorders [188,189].

The idea that misfolded proteins can be rescued comes true because several strategies, including chaperone overexpression or using proteasome inhibitors, facilitate the biogenesis and surface expression of these wasteful proteins, indicating that they were viable folding intermediates able to integrate into functional pools [190].

5. Aid for Misfolded Proteins

In 1978, Ronald Laskey and his colleagues were the first to describe a chaperone protein necessary for nucleosome assembly (nucleoplasmin) [191]. Chaperones can be expressed either in different regions or in a tissue-specific manner, and the target proteins binding to chaperones are usually termed clients [192]. Even when the protein structure is determined through the amino acid sequence, chaperones are essential regulators stabilizing protein structure and participating in the folding and assembly processes. Chaperones can bind to nascent or unfolded proteins to stabilize their structure and prevent the formation of aggregates within the cell [193–195]. To date, different kinds of proteins and small molecules that can be used physiologically to facilitate protein folding and assembly have been identified. Chaperones are formally classified as molecular, chemical, and pharmacological chaperones according to their origin [196,197].

5.1. Molecular Chaperones Associated with Plasma Membrane Protein Biogenesis

Molecular chaperones are a family of proteins that can recognize conflicts among paired-contact interactions in proteins, where the implicated amino acids cannot reach a minimal free energy state while folding (i.e., a frustrated structural state) [198]. Most of these chaperones were discovered as proteins expressed in response to temperature changes or environmental stresses and were named heat shock proteins (Hsp) [199].

Chaperones can work as holdases (holding a partial folding state of a protein), foldases (assisting protein folding in an ATP-dependent manner), or unfoldases (unfolding transiently unfolded intermediates to allow protein refolding) [200,201]. The mechanism by which chaperones recognize and assist their client proteins is still not clear as most chaperones are promiscuous molecules. Although some interactions can be regulated in an ATP-dependent manner, others cannot, suggesting that the molecular mechanism of chaperones may not be unique.

Studies have demonstrated the interaction of chaperones with polypeptides during folding—through hydrophobic amino acid residues—to avoid aggregate formation and stabilize the structure [202,203]; however, the way chaperones interact with native folded proteins is not clear. A computational high-throughput analysis of the Hsp90 client (an ATP-dependent chaperone) failed to identify a consensus motif allowing interaction [204]. Recently, researchers solved the interaction of bacterial chaperones (e.g., ATP-independent holdases Spy, SurA, and Skp, and the ATP-dependent chaperone GroEL) with clients, demonstrating that chaperone–client interaction occurs regardless of sequence motif or local structure, but binding happens locally through an entropy-based mechanism when a protein surface is frustrated [205,206] and unable to achieve a minimum energy structural state [207,208]. Researchers have suggested several transient local interactions with some segments of the client during folding to reach a minimally frustrated conformation [198].

Hsp function is crucial because these proteins promote cell stability under stress or pathological conditions. Cells would suffer irreversible damage or even cell death without Hsps [209]. Twenty different protein families exhibit chaperone activity. These proteins are classified by their molecular weight and offer different alternatives for correct folding. To date, just a few molecular chaperones have been associated with PM biogenesis.

5.1.1. Hsp70

Binding immunoglobulin protein (BiP), a member of the heat shock protein 70 (Hsp70) family, has been found in the ER lumen, where it plays a role in the recognition of misfolded MPs. Hsp70 folding is assisted by other chaperones such as Hsp90 and Hsp40, DNAJA1, and DNAJB1. Hsp70 unfolds proteins through ATP hydrolysis, inducing the cyclic binding and releasing of hydrophobic amino acids [210,211].

The balance between ER export and retention mediated through Hsp70–Hsp90 cytosolic chaperone systems [189,212] has been evident in the trafficking of mutant ion channels with impaired traffic, such as voltage-gated delayed rectifier potassium channel, hERG (human ether-a-go-go-related gene), which is related to congenital long QT syndrome type 2 (LQT2) [213], and cystic fibrosis transmembrane conductance regulator (CFTR), a chloride channel that plays an important role in the maintenance of ion balance. CFTR favors epithelial surface hydration prominently in the lung airways and pancreas, and its dysfunction is implicated in cystic fibrosis [189,214].

In vitro co-expression of Hsp70 or Hsc70 (heat shock cognate protein, a constitutively expressed member of the Hsp70 family) with mutant hERG or voltage-sensitive potassium channel Kv1.5 prolonged the channel lifetime and increased functionality at the cell surface, decreasing its ubiquitination [215]. However, overexpression of Hsp70 with the co-chaperone DNAJB1 or Hsp73 only induced modest traffic and stabilization improvement of the deletion of phenylalanine at position 508) in the CFTR channel (ΔF508-CFTR) [216,217]. Additionally, Hsc70 and one of its co-chaperones, DNAJA1, are more associated with ΔF508-CFTR than with CFTR wild-type, suggesting that chaperones engage in a mutant channel trying to refold it [218]. Mixed tetrameric channels integrated by acid-sensing ion channel subunits (ASIC1 or ASIC2) together with subunits forming an epithelial sodium channel (ENaCα or ENaCγ) are thought to form the active ASIC channels at the glioma cell surface [219,220]. Hsc70 was found to associate with ASIC2 in glioma cells [221]. Knocking down this chaperone reduced the channel activity but increased the cell surface expression and inhibited glioma cell migration [219].

Thus, these chaperones promote protein damage recovery and improve cell viability. From another functional perspective, overexpression of Hsp70 also suppresses phenotypes related to protein aggregation in models of Alzheimer's disease [222] and Parkinson's disease [223–225].

5.1.2. Hsp90

The family of Hsp90 chaperones comprises critically conserved proteins that are major molecular chaperones within eukaryotic cells. Hsp90 proteins are important for cellular stabilization processes involving signal transduction, cellular trafficking, chromatin remodeling, cell growth, differentiation, and reproduction [226–230]. Hsp90 interacts with specific proteins through an ATP-dependent cycle, guaranteeing folding, transport, and/or assembly of the client into multiprotein complexes [231,232]. In addition to assisting in the folding of nascent proteins, Hsp90 chaperones promote protein refolding and aberrant protein degradation, possibly indicating that Hsp90 proteins can adapt their conformation to match every client or that they recognize different clients in different conformations [233–236].

Hsp90 client proteins include transcription factors (e.g., HIF1α, ATF3, and p53), steroid hormone receptors (e.g., estrogen, glucocorticoid, and progesterone receptors), and kinases (e.g., EGFR, B-raf, and SRC), among other proteins. Many of these client proteins are commonly overexpressed and/or frequently mutated in cancer cells [237–239].

Early studies identified Hsp90 in complexes with hERG or CFTR channels. Hsp90 inhibition impaired hERG trafficking, but the effect on CFTR remains unknown [240].

Hsp90β overexpression restored PM expression of the mutated voltage-gated potassium channel (KCNQ4) related to autosomal dominant deafness type 2A and increased the activity of the WT channel, but not the activity of WT and mutant (W276S) mixed channels, suggesting that chaperone affinity is affected due to the mutation [241].

Since Hsp90 regulates the stability of oncoproteins important in tumor development and progression, in addition to controlling other pathological oligomeric aggregates causing neurodegenerative diseases, Hsp90 inhibitors have been suggested as a potential therapy for misfolding-related diseases [242,243].

Under proteotoxic stress conditions, or when cellular protein degradation is overwhelmed, misfolded MPs can become stuck and intracellularly form aggresomes [244], which should be eventually cleared through autophagy over kinetically controlled chaperone-assisted folding and degradation systems [244,245], including a group of molecular chaperones working together with other molecules with chaperone activity (e.g., calnexin and a protein disulfide isomerase that catalyzes the formation of disulfide bonds, allowing proteins to fold) [246–249].

How misassembled transmembrane domains are recognized is not clearly understood. However, because protein maturation is a complex procedure, it is expected that a recently proposed conformational frustration state recognized by chaperones [198,207,208] may consider intrinsic structural properties of the misfolded proteins [89,250]; for instance, single amino acid mutations within protein domains could potentially disrupt native helical packing interactions inducing solvation of TM segments that are naturally hydrophobic [251,252], thus prompting the formation of different protein contacts or interactions affecting their structure.

BiP activity was experimentally evidenced for two proteins needing a β subunit to stabilize their membrane location: the Na$^+$/K$^+$ ATPase [251,253,254] and the single-pass TM α subunit of the αβ T-cell receptor (αβTCR) [255]. For both proteins, the lack of a β subunit left some residues of the α subunit out of the membrane, thus favoring protein interaction with BiP—targeting proteins for degradation—and reducing the probability of exporting immature proteins.

5.1.3. Co-Chaperones Cooperating in Membrane Protein Folding

Cooperation between molecular chaperones to control protein synthesis and degradation is crucial for cell maintenance. The physiological role of chaperones is evident because proteins are not always able to attain their correct conformation spontaneously, and some factors, such as age, disease, or cellular stress, may also influence protein folding, thus affecting the balance of protein "rescue" or protein degradation. Many co-chaperones have been reported to regulate chaperone activity related to degradation. Particularly, a highly conserved cytoplasmic protein called CHIP (carboxyl terminus of Hsc70 interacting protein) was identified during protein screening, in addition to the tetratricopeptide repeat (TPR) [256], a very conserved domain in several co-chaperones. TRP allows CHIP interaction with Hsp70, Hsc70, or Hsp90 to induce client substrate ubiquitylation and proteasome degradation because CHIP works as a ubiquitin ligase [257].

Some studies showed that CHIP-dependent polyubiquitination serves as a sorting signal for internalization and lysosomal degradation of the MP, including mutants from dopamine receptor D4.4, vasopressin V2 receptor (V2R), CFTR and G-protein coupled receptors (GPCRs) [258]. Later on, a supportive study showed that in vitro co-expression of CHIP with voltage-gated potassium channel Kv1.5 increased channel ubiquitination and decreased the protein level, while CHIP suppression increased channel expression [259].

Many mutations in genes encoding structural proteins may not influence protein functionality but may still cause some diseases by either arresting mutant proteins intracellularly or preventing the cellular trafficking machinery from transporting the mutant protein to an appropriate subcellular location to avoid protein misfolding and aggregate formation in the first step of the pathological cascade. Considering that chaperones are essential to many physiological processes by preventing protein misfolding and aggregation, several strategies based on their buffering capacity are rapidly emerging as promising treatments for misfolding-related diseases [196,197].

The ability to restore the biosynthesis of misfolded proteins has been demonstrated through different strategies. For instance, ΔF508-CFTR channel is a classic example of a mutation causing a misfolding disease. Nevertheless, crystal structures and biophysical studies comparing WT and mutant CFTR domains suggest only local structural changes due to the amino acidic deletion [260,261]. Several studies have demonstrated that F508 deletion induces ER retention, proteolytic degradation, and absent Cl⁻ conductance. The ability of CFTR to traffic to the PM at 37 and <30 °C was evaluated, and the lower temperature was found to favor PM localization and function [262]. Many other chaperones have been evaluated as potent modulators of several misfolding diseases.

5.2. Chemical Chaperones

Some small organic molecules helping to maintain proper proteostasis in stressful environments are considered chemical chaperones because they enhance the folding and/or stability of proteins. These molecules can be present in a diverse range of organisms or tissues under denaturing conditions [263] and are potentially helpful in treating conformational diseases. Nonspecific interactions with unrelated proteins make them not specific for a therapeutic target. Some examples of these compounds are polyols, such as glycerol, dimethyl sulfoxide (DMSO), trimethylamine (trimethylamine N-oxide [TMAO]) and amino acid derivatives, 4-phenylbutyric acid (4PBA), membrane-permeable forms of promiscuous enzyme antagonists, ligands, and substrates. Thus far, the mechanism by which chemical chaperones modulate folding energy landscapes is not clearly understood [264], but these chaperones stabilize misfolded proteins, thus decreasing the formation of protein aggregates, preventing unnecessary interactions with other proteins and altering the activity of other chaperones such that proteins are transported to their final destination with increased efficiency. Chemical chaperones of the polyol TMAO and 4PBA groups act on multiple proteins, whereas antagonists, ligands, and substrates affect specific proteins (and are commonly called pharmacological chaperones) [265].

Chemical chaperones can be classified into two groups: osmolytes and hydrophobic compounds. Even when they lack specificity, these molecules usually have an effect only at high concentrations and are thus frequently rejected as therapeutic agents even when they rescue the misfolded protein state in vitro [190,261]. De novo design of chemical chaperones with increased activity and specificity is desirable to ameliorate protein misfolding and aggregation in different contexts.

Osmolytes: The group of osmolyte chaperones comprises low molecular weight compounds, including free amino acids and amino acid derivatives (glycine, taurine, and β-alanine). Other osmolyte compounds are polyols, such as glycerol and sucrose, and methylamines, particularly TMAO. Osmolyte chaperones are crucial for organisms exposed to stressful conditions such as fluctuating salinity, desiccation, or extreme temperatures [266]. These types of chaperones increase protein stability without disrupting protein function. Polyols protect cells against extreme conditions such as increased temperature or dehydration. Amino acids preserve proteins in high-salinity environments, and methylamines protect cells against the denaturing effect of urea [266,267].

Although osmolyte chaperones have effects on different kinds of proteins and under diverse conditions, they share the function of stabilizing protein structure by modifying the solvent properties. Osmolyte chaperone solvation decreases water activity around each polypeptide, forcing partially exposed hydrophobic patches to reach the most stable conformation and facilitate protein folding [268–270].

Several studies have shown that small organic compounds which stabilize PM protein [98] can rescue protein folding defects by increasing traffic and function at the PM for selective mutants on the cystic fibrosis-related CFTR chloride channel [98,271], aquaporin-2 water channel (AQP2), and V2R associated with nephrogenic diabetes insipidus (NDI) [272,273]. Some of the mutants in these studies were also tested with either TMAO, DMSO, or glycerol, showing different rescuing effects. TMAO showed a lower ability to rescue MPs, which may be due to an enhanced hydration potential that affects the reagent permeability on lipidic membranes or the stability of hydrophobic regions on the MP [274,275].

Hydrophobic compounds: Several molecules have been classified as hydrophobic chaperones, including (4PBA) and bile acids (e.g., oxysterol, an oxygenated isoform of cholesterol).

4PBA is capable of restoring the cell surface expression of mislocated mutants on bile salt export pump (BSEP), which is related to an inherited autosomal recessive liver disease called progressive familial intrahepatic cholestasis type 2 (PFIC2) that leads to cirrhosis and death before adulthood [276]. In a recent study testing a multidrug regimen of 4PBA mixed with anticonvulsants oxcarbazepine and maralixibat, PFIC2 symptoms were controlled in two siblings with partial loss of BSEP activity [277]. 4PBA has also been tested with the mutant channel Δ508-CFTR-inducing cell surface expression [278], and some clinical trials using this chemical chaperone in cystic fibrosis patients have demonstrated an improvement in CFTR function in the nasal epithelia [279].

For cyclic nucleotide-gated channels related to retinopathies, reduced degradation and/or promoted PM localization of defective subunits was shown using chemical chaperones such as 4PBA or the bile acid component tauroursodeoxycholic acid (TUDCA) [280].

The general mechanism of action proposed for hydrophobic chaperones is based on the interaction between the chaperone's hydrophobic regions and the exposed hydrophobic segments of the unfolded protein. However, even though 4PBA and bile acids can reduce aggregate accumulation in vivo and in vitro and revert ER stress, these molecules may be more complex in the action mechanism influencing different levels of regulation [196].

5.3. Pharmacological Chaperones

Pharmacological chaperones, also known as pharmacoperones, are lipophilic compounds stabilizing protein conformation to prevent degradation and promote proper trafficking to their functional site of action in the cell. These molecules represent one of the most promising therapeutic strategies to treat misfolding-related diseases [281]. Unlike chemical chaperones, these low molecular weight compounds bind selectively to proteins, stabilizing the protein structure and restoring protein localization and function. Enzymes, agonists, antagonists, and some synthetic compounds are examples of pharmacoperones [282].

Antagonists used as chaperones should ideally provide an effective rescue response at the lowest concentration, without losing the ability to dissociate from the rescued protein to facilitate the endogenous ligand binding [283]. Antagonists are highly efficacious in rescuing mislocated mutant proteins by preventing agonist or substrate access [284–286]; however, antagonists of high affinity for receptors and ion channels may yield nonfunctional channels expressed at the PM, as was shown for the rescued ATP-sensitive potassium channels (KATP) treated with sulfonylureas [287].

Agonist molecules were identified as pharmacoperones when SR49059, a small, cell-permeable molecule formerly developed as a vasopressin antagonist, was able to rescue the function of ER-retained V2 vasopressin receptor (V2R) mutants [288,289]. The experiments clearly showed significantly improved kidney function in NDI patients [289]. Rhodopsin-like G-protein-coupled MC4R (melanocortin 4 receptor), which causes severe early-onset morbid obesity in humans, exemplifies misfolded and intracellularly retained proteins rescued using antagonists [290].

Channel blockers also represent an alternative to modulate protein folding and trafficking; cisapride, E-4031, and the non-specific antiarrhythmic drug quinidine have experimentally rescued mutant hERG leading to LQT2. However, using channel blocker as a pharmacological strategy is still controversial as blocking ion flow can affect K^+ flux and cell homeostasis, leading to a prolonged QT interval and increased risk of developing arrhythmia [291–293].

The dissociation rate is not essential for using an agonist as a chaperone [294,295]. Mislocation of mutant MPs has been rescued using agonists. Defective trafficking of misfolded mutant CNG channels has been successfully rescued using a cell-permeable cyclic nucleotide agonist [280]. High-affinity non-peptide agonists were able to selectively rescue PM expression and function of misfolded arginine-vasopressin receptor 2 (AVPR2) mutants associated with NDI [296]. Some mutants in the calcium-sensing receptor (CaSR) leading to familial hypocalciuric hypercalcemia and neonatal hyperparathyroidism can also be rescued using membrane-permeant allosteric agonists to recover protein functionality [190,295].

In addition to functioning as cell proteostasis modulators modifying the cell proteome and increasing MP maturation, pharmacological chaperones can also act as correctors and potentiators. In fact, for the CFTR mutant channel, it has been tested that some compounds, such as corr-4a and VRT-532, interact directly with the misfolded protein to correct the biosynthetic pathway and enhance trafficking and channel function [297]. VX-809 (Lumacaftor) is a corrector tested by itself or in combination with other molecules [298] in patients with cystic fibrosis and is a potential treatment for other pathologies associated with misfolding and misrouted proteins, as has been proved to Stargardt disease, which invariably ends in legal blindness (visual acuity of 20/200 or less [299]) [298]. Among other correctors or pharmacoperones, Lumacaftor represents hope to set up therapeutic strategies for misfolding diseases [300,301].

Assistance to stabilize protein folding can be analyzed in vivo or in vitro, as proteins can naturally interact with many compounds intracellularly. Proteins can be assisted by chemical chaperones, which helps scientists to better understand how proteins are properly folded in vitro and thus find more suitable and specific compounds (or pharmacological chaperones) to develop a therapeutic strategy for misfolding diseases.

Table 2 summarizes the misfolded PM proteins related to pathologies and the therapeutic approaches used; molecular, chemical, and pharmacological chaperones are widely regarded as a promising therapeutic strategy for these pathologies.

Table 2. Chaperones used to rescue misfolded plasma membrane proteins related to diseases.

Misfolded Membrane Protein	Disease	Gene	Rescuing Strategy		
			Molecular Chaperones	Chemical Chaperones	Pharmacological Chaperones
α-synuclein	Parkinson's disease	SNCA	Hsp70 [223–225]		
Aquaporin-2	Autosomal Nephrogenic Diabetes Insipidus	AQP2		Glycerol, Trimethylamine-N-oxide (TMAO) and Dimethyl sulfoxide (DMSO) [2,272,273].	
Arginine-Vasopressin (AVP) Receptor 2 (AVPR$_2$)	Nephrogenic Syndrome of Inappropriate Antidiuresis and Diabetes Insipidus (nephrogenic, X-Linked)	AVPR2			OPC51803, VA999088, and VA999089 [293]. L44P, A294P, and R337X [294]. SR49059, VPA-985, OPC31260, OPC41061 (Tolvaptan) and SR121463B [293,302,303]. MCF14, MCF18, and MCF57 [294,304].
ATP-binding Cassette	Tangier disease	ABCA1		Sodium 4-Phenylbutyrate (4-PBA) [305].	
	Stargardt Eye disease	ABCA4			VX-809 (Lumacaftor) [306].
Bile Salt Export Pump (BSEP)	Progressive Familial Intrahepatic Cholestasis type 2	ABCB11		4-PBA mixed with Anticonvulsant-Oxcarbazepine, and Maralixibat [274,275].	
Calcium-Sensing Receptor (CaSR)	Familial Hypocalciuric Hypercalcemia	CaSR			MG132, NPS R-568 [295,307].
Cardiac Sodium (Na$^+$) Channel NaV1.5	Brugada Syndrome Nocturnal Death syndrome	SCN5A		Curcumin [308].	Mexiletine [309].
Connexin Cx31, Cx43, Cx50	Charcot-Marie-Tooth syndrome	GJA1			Cycloheximide [310].
Copper-transporting P-type ATPase	Menkes disease	ATP7A		Excess of copper [311].	Copper Toxicosis Protein COMMD1 [311].
Cyclic Nucleotide Gated (CNG) Channel	Retinitis Pigmentosa, Achromatopsia	CNGA3		TUDCA (Tauroursodeoxycholate Sodium salt), 4-PBA [278]. Glycerol [312]. MTSHB (Hydroxybenzyl-Methanethiosulfonate), MTSEA (Aminoethyl-Methanethiosulfonate) [123].	CPT-cGMP [8-(chlorophenylthio)-cGMP] [278].
Cystic Fibrosis Transmembrane Conductance Regulator (CFTR)	Cystic Fibrosis	CFTR	Hsc70, Hsp90 [313]. Hsc70/Hdj-2 [218].	Glycerol [98]. TMAO [269]. 4-PBA [276,277].	VX-809 (Lumacaftor) [269]. Lumacaftor/Ivacaftor [314,315]. Cycloheximide [316]. Corr-4a and VRT-532 [317,318]. VX-661(Tezacaftor)/Ivacaftor [319].
Dopamine Transporter (DAT)	Infantile parkinsonism-dystonia	SLC6A3			Ibogaine, Noribogaine [320–322].
Gonadotropin Releasing Hormone Receptor (GnRHR)	Hypogonadotropic hypogonadism	GNRHR	JB12, Hsp70 [323].		IN3 [324].
HERG potassium channel	Hereditary long QT syndrome	KCNH2	sp40/DnaJ [325].		E-4031 [289,326]. Cisapride [291]. Thapsigargin [290].
Insulin receptor	Diabetes Mellitus, Insulin-resistant syndrome	INSR	Calnexin and Calreticulin [327].		
Melanocortin-4 receptor (MC4R)	Severe early-onset morbid obesity	MC4R			Ipsen 17 [288,328]. ML00253764, Ipsen 5i [328–330].
Voltage-gated potassium channel (VGKC)	Autosomal Dominant Deafness type 2A	KCNQ4	Hsp90β [241].		
Neuroligin-3	X-linked autism, Asperger syndrome	NLGN3	Calnexin [331].		
Pendrin	Pendred syndrome and Non-syndromic Hearing loss	SLC26A4 (PDS)		TMAO [332].	Cycloheximide (CHX), Puromycin [332].
Prion Protein (PrP)	Genetic Creutzfeldt-Jakob disease, Gerstmann Straussler Scheinker syndome and Fatal Familial Insomnia	PRNP	BiP [333], [334].		
Rhodopsin	Retinitis Pigmentosa	RHO	1-*cis* retinal [335].	DMSO [336]. Curcumin [337].	YC-001 [338]. S-RS1 [336].
Sodium-borate cotransporter	Corneal dystrophy	SLC4A11			Anti-inflammatory drugs (NSAIDs), Glafenine, Ibuprofen, and Acetylsalicylic acid dissolved in DMSO [339].
Vasopressin Type 2 Receptor (V2R)	Nephrogenic Diabetes Insipidus	V2R		Glycerol, DMSO and TMAO [272,273].	SR49059 [197,286,287]. Thapsigargin/Curcumin and Ionomycin [340].
Voltage sensitive potassium channel (Kv1.5)	Atrial Fibrillation	KCNA5	Hsp70 [215].		

6. Conclusions

MP localization is critical for normal cell physiology. The correct routing of proteins is as important as the genetic machinery for protein expression. Protein trafficking, moreover, is not a simple or unidirectional process. Naturally, trafficking assistance from the ER to the plasma membrane involves many proteins and factors as protein carriers or chaperones of molecular cargo, and, as a response to misfolding, chaperones may also mask harmful modifications in the folding energy landscape. However, when protein folding proceeds incorrectly, many diseases can arise; a wide range of diseases result when protein misfolding induces intracellular retention of MPs. Misassembled proteins that reach the membrane can also lead to disease due to dysfunctionality.

Biomolecules **2020**, *10*, 728

Currently, specific therapies for conformational diseases are lacking because of a gap in the understanding of the mechanisms by which the natural conformation of proteins is altered into many misfolded pathological forms. However, encouragingly, an extensive and concerted effort is already underway to combat misfolding diseases. The discovery of compounds with therapeutic chaperone ability is customarily initiated through high-throughput screening of libraries, including natural or synthesized compounds, searching for stabilizing binders using in silico, in vitro, in vivo, or cell-based approaches. As we continue increasing knowledge of disease mechanisms, we also continue to discover molecules that could interact with PM proteins, mimicking the effect of natural chaperones that correct misfolding.

Author Contributions: K.J.-N. and A.L.-R. conceptualized the idea and wrote the original draf; V.M.A.-G., E.R.-B., I.M.-M., J.L.R.-B. made substantial contributions to conception, design, and/or acquisition of data; A.L.-R. prepared the figures. Funding acquisition by A.L.-R. All authors have read and agreed to the published version of the manuscript.

Funding: This research was funded by UJED-CONACYT-CB-2015-01-259091. K.J.-N., and J.L.R.-B., were supported by CONACYT 637362 and 638336, respectively.

Acknowledgments: The authors would like to thank the support of UJED-CONACYT-CB-2015-01-259091. K.J.-N. and J.L.R.-B. were supported by CONACYT 637362 and 638336, respectively. We give special thanks to Ataulfo Martínez-Torres, Jessica, G.; Norris, Hugo, R. Masse-Torres, and Valeria Amaya-Galnarez for reading and commenting on this manuscript and to the members of the NFMyC lab, Adriana, R., Angeles, A., Sofía, N., and Yuvia, C., for contributing with comments, data, and discussions.

Conflicts of Interest: The authors declare no conflict of interest.

References

1. Muro, E.; Atilla-Gokcumen, G.E.; Eggert, U.S. Lipids in cell biology: How can we understand them better? *Mol. Biol. Cell* **2014**, *25*, 1819–1823. [CrossRef] [PubMed]
2. O'Brien, J.S. Cell membranes—Composition: Structure: Function. *J. Theor. Biol.* **1967**, *15*, 307–324. [CrossRef]
3. Engelman, D.M. Membranes are more mosaic than fluid. *Nature* **2005**, *438*, 578–580. [CrossRef] [PubMed]
4. Deisenhofer, J.; Epp, O.; Miki, K.; Huber, R.; Michel, H. Structure of the protein subunits in the photosynthetic reaction centre of Rhodopseudomonas viridis at 3Å resolution. *Nature* **1985**, *318*, 618–624. [CrossRef]
5. Block, R.C.; Dorsey, E.R.; Beck, C.A.; Brenna, J.T.; Shoulson, I. Altered cholesterol and fatty acid metabolism in Huntington disease. *J. Clin. Lipidol.* **2010**, *4*, 17–23. [CrossRef]
6. Di Paolo, G.; Kim, T.-W. Linking lipids to Alzheimer's disease: Cholesterol and beyond. *Nat. Rev. Neurosci.* **2011**, *12*, 284–296. [CrossRef]
7. Lee, S.; Zheng, H.; Shi, L.; Jiang, Q.-X. Reconstitution of a Kv channel into lipid membranes for structural and functional studies. *J. Vis. Exp.* **2013**, e50436. [CrossRef]
8. Jiang, Q.-X. Cholesterol-dependent gating effects on ion channels. *Adv. Exp. Med. Biol.* **2019**, *1115*, 167–190. [CrossRef]
9. Finol-Urdaneta, R.K.; McArthur, J.R.; Juranka, P.F.; French, R.J.; Morris, C.E. Modulation of KvAP unitary conductance and gating by 1-alkanols and other surface active agents. *Biophys. J.* **2010**, *98*, 762–772. [CrossRef]
10. Balijepalli, R.C.; Delisle, B.P.; Balijepalli, S.Y.; Foell, J.D.; Slind, J.K.; Kamp, T.J.; January, C.T. Kv11.1 (ERG1) K+ channels localize in cholesterol and sphingolipid enriched membranes and are modulated by membrane cholesterol. *Channels* **2007**, *1*, 263–272. [CrossRef]
11. Abi-Char, J.; Maguy, A.; Coulombe, A.; Balse, E.; Ratajczak, P.; Samuel, J.-L.; Nattel, S.; Hatem, S.N. Membrane cholesterol modulates Kv1.5 potassium channel distribution and function in rat cardiomyocytes. *J. Physiol.* **2007**, *582*, 1205–1217. [CrossRef] [PubMed]
12. Poveda, J.A.; Giudici, A.M.; Renart, M.L.; Millet, O.; Morales, A.; González-Ros, J.M.; Oakes, V.; Furini, S.; Domene, C. Modulation of the potassium channel KcsA by anionic phospholipids: Role of arginines at the non-annular lipid binding sites. *Biochim. Biophys. Acta Biomembr.* **2019**, *1861*, 183029. [CrossRef] [PubMed]
13. Milescu, M.; Vobecky, J.; Roh, S.H.; Kim, S.H.; Jung, H.J.; Kim, J., II; Swartz, K.J. Tarantula toxins interact with voltage sensors within lipid membranes. *J. Gen. Physiol.* **2007**, *130*, 497–511. [CrossRef]
14. Kim, R.Y.; Pless, S.A.; Kurata, H.T. PIP2 mediates functional coupling and pharmacology of neuronal KCNQ channels. *Proc. Natl. Acad. Sci. USA* **2017**, *114*, E9702–E9711. [CrossRef] [PubMed]

15. Delgado-Ramírez, M.; López-Izquierdo, A.; Rodríguez-Menchaca, A.A. Dual regulation of hEAG1 channels by phosphatidylinositol 4,5-bisphosphate. *Biochem. Biophys. Res. Commun.* **2018**, *503*, 2531–2535. [CrossRef]

16. Bissig, C.; Gruenberg, J. Lipid sorting and multivesicular endosome biogenesis. *Cold Spring Harb. Perspect. Biol.* **2013**, *5*, a016816. [CrossRef]

17. Huang, C.-L. Complex roles of PIP2 in the regulation of ion channels and transporters. *Am. J. Physiol. Renal Physiol.* **2007**, *293*, F1761–F1765. [CrossRef]

18. Thapa, N.; Sun, Y.; Schramp, M.; Choi, S.; Ling, K.; Anderson, R.A. Phosphoinositide signaling regulates the exocyst complex and polarized integrin trafficking in directionally migrating cells. *Dev. Cell* **2012**, *22*, 116–130. [CrossRef]

19. Hernández-Araiza, I.; Morales-Lázaro, S.L.; Canul-Sánchez, J.A.; Islas, L.D.; Rosenbaum, T. Role of lysophosphatidic acid in ion channel function and disease. *J. Neurophysiol.* **2018**, *120*, 1198–1211. [CrossRef]

20. Bukiya, A.N.; Dopico, A.M. Regulation of BK Channel activity by cholesterol and its derivatives. *Adv. Exp. Med. Biol.* **2019**, *1115*, 53–75. [CrossRef]

21. Stevens, T.J.; Arkin, I.T. Do more complex organisms have a greater proportion of membrane proteins in their genomes? *Proteins* **2000**, *39*, 417–420. [CrossRef]

22. von Heijne, G.; Gavel, Y. Topogenic signals in integral membrane proteins. *Eur. J. Biochem.* **1988**, *174*, 671–678. [CrossRef] [PubMed]

23. Haynes, C.M.; Petrova, K.; Benedetti, C.; Yang, Y.; Ron, D. ClpP Mediates activation of a mitochondrial unfolded protein response in C. elegans. *Dev. Cell* **2007**, *13*, 467–480. [CrossRef] [PubMed]

24. Ron, D.; Walter, P. Signal integration in the endoplasmic reticulum unfolded protein response. *Nat. Rev. Mol. Cell Biol.* **2007**, *8*, 519–529. [CrossRef]

25. Sardiello, M.; Palmieri, M.; di Ronza, A.; Medina, D.L.; Valenza, M.; Gennarino, V.A.; Di Malta, C.; Donaudy, F.; Embrione, V.; Polishchuk, R.S.; et al. A gene network regulating lysosomal biogenesis and function. *Science* **2009**, *325*, 473–477. [CrossRef]

26. Johnson, A.E.; van Waes, M.A. The Translocon: A Dynamic gateway at the ER membrane. *Annu. Rev. Cell Dev. Biol.* **1999**, *15*, 799–842. [CrossRef]

27. Stefani, M.; Rigacci, S. Protein folding and aggregation into amyloid: The interference by natural phenolic compounds. *Int. J. Mol. Sci.* **2013**, *14*, 12411–12457. [CrossRef]

28. Singer, S.J. The Structure and insertion of integral proteins in membranes. *Annu. Rev. Cell Biol.* **1990**, *6*, 247–296. [CrossRef]

29. Sabatini, D.D.; Kreibich, G.; Morimoto, T.; Adesnik, M. Mechanisms for the incorporation of proteins in membranes and organelles. *J. Cell Biol.* **1982**, *92*, 1–22. [CrossRef]

30. Blobel, G. Protein Targeting (Nobel Lecture). *ChemBioChem* **2000**, *1*, 86–102. [CrossRef]

31. Peer, W.A. *Plasma Membrane Protein Trafficking BT—The Plant Plasma Membrane*; Murphy, A.S., Schulz, B., Peer, W., Eds.; Springer: Berlin/Heidelberg, Germany, 2011; pp. 31–56, ISBN 978-3-642-13431-9.

32. Morozova, D.; Guigas, G.; Weiss, M. Dynamic structure formation of peripheral membrane proteins. *PLoS Comput. Biol.* **2011**, *7*, e1002067. [CrossRef] [PubMed]

33. Hayer, A.; Stoeber, M.; Bissig, C.; Helenius, A. Biogenesis of caveolae: Stepwise assembly of large caveolin and cavin complexes. *Traffic* **2010**, *11*, 361–382. [CrossRef]

34. Monier, S.; Dietzen, D.J.; Hastings, W.R.; Lublin, D.M.; Kurzchalia, T. V Oligomerization of VIP21-caveolin in vitro is stabilized by long chain fatty acylation or cholesterol. *FEBS Lett.* **1996**, *388*, 143–149. [CrossRef]

35. Busija, A.R.; Patel, H.H.; Insel, P.A. Caveolins and cavins in the trafficking, maturation, and degradation of caveolae: Implications for cell physiology. *Am. J. Physiol. Cell Physiol.* **2017**, *312*, C459–C477. [CrossRef] [PubMed]

36. Pestova, T.V.; Kolupaeva, V.G.; Lomakin, I.B.; Pilipenko, E.V.; Shatsky, I.N.; Agol, V.I.; Hellen, C.U. Molecular mechanisms of translation initiation in eukaryotes. *Proc. Natl. Acad. Sci. USA* **2001**, *98*, 7029–7036. [CrossRef] [PubMed]

37. Hegde, R.S.; Keenan, R.J. Tail-anchored membrane protein insertion into the endoplasmic reticulum. *Nat. Rev. Mol. Cell Biol.* **2011**, *12*, 787–798. [CrossRef]

38. Kapp, K.; Schrempf, S.; Lemberg, M.; Dobberstein, B. Post-targeting functions of signal peptides. In *Protein Transport into the Endoplasmic Reticulum*; Landes Bioscience: Austin, TX, USA, 2000.

39. Liu, X.; Zheng, X.F.S. Endoplasmic reticulum and golgi localization sequences for mammalian target of rapamycin. *Mol. Biol. Cell* **2007**, *18*, 1073–1082. [CrossRef]

40. Frangioni, J.V.; Beahm, P.H.; Shifrin, V.; Jost, C.A.; Neel, B.G. The nontransmembrane tyrosine phosphatase PTP-1B localizes to the endoplasmic reticulum via its 35 amino acid C-terminal sequence. *Cell* **1992**, *68*, 545–560. [CrossRef]

41. Pottekat, A.; Menon, A.K. Subcellular localization and targeting of N-acetylglucosaminyl phosphatidylinositol de-N-acetylase, the second enzyme in the glycosylphosphatidylinositol biosynthetic pathway. *J. Biol. Chem.* **2004**, *279*, 15743–15751. [CrossRef]

42. Ercan, E.; Momburg, F.; Engel, U.; Temmerman, K.; Nickel, W.; Seedorf, M. A conserved, lipid-mediated sorting mechanism of yeast Ist2 and mammalian STIM proteins to the peripheral ER. *Traffic* **2009**, *10*, 1802–1818. [CrossRef]

43. Sun, Q.; Ju, T.; Cummings, R.D. The transmembrane domain of the molecular chaperone Cosmc directs its localization to the endoplasmic reticulum. *J. Biol. Chem.* **2011**, *286*, 11529–11542. [CrossRef] [PubMed]

44. Haugsten, E.M.; Malecki, J.; Bjørklund, S.M.S.; Olsnes, S.; Wesche, J. Ubiquitination of fibroblast growth factor receptor 1 is required for its intracellular sorting but not for its endocytosis. *Mol. Biol. Cell* **2008**, *19*, 3390–3403. [CrossRef] [PubMed]

45. Watanabe, K.; Nagaoka, T.; Strizzi, L.; Mancino, M.; Gonzales, M.; Bianco, C.; Salomon, D.S. Characterization of the glycosylphosphatidylinositol-anchor signal sequence of human Cryptic with a hydrophilic extension. *Biochim. Biophys. Acta* **2008**, *1778*, 2671–2681. [CrossRef] [PubMed]

46. Lanza, F.; De La Salle, C.; Baas, M.-J.; Schwartz, A.; Boval, B.; Cazenave, J.-P.; Caen, J.P. A Leu7Pro mutation in the signal peptide of platelet glycoprotein (GP)IX in a case of Bernard-Soulier syndrome abolishes surface expression of the GPIb-V-IX complex. *Br. J. Haematol.* **2002**, *118*, 260–266. [CrossRef]

47. Luo, W.; Marsh-Armstrong, N.; Rattner, A.; Nathans, J. An outer segment localization signal at the C terminus of the photoreceptor-specific retinol dehydrogenase. *J. Neurosci.* **2004**, *24*, 2623–2632. [CrossRef]

48. Klapisz, E.; Ziari, M.; Wendum, D.; Koumanov, K.; Brachet-Ducos, C.; Olivier, J.L.; Béréziat, G.; Trugnan, G.; Masliah, J. N-terminal and C-terminal plasma membrane anchoring modulate differently agonist-induced activation of cytosolic phospholipase A2. *Eur. J. Biochem.* **1999**, *265*, 957–966. [CrossRef]

49. Zlatkine, P.; Mehul, B.; Magee, A.I. Retargeting of cytosolic proteins to the plasma membrane by the Lck protein tyrosine kinase dual acylation motif. *J. Cell Sci.* **1997**, *110 Pt 5*, 673–679.

50. Beau, I.; Groyer-Picard, M.-T.; Desroches, A.; Condamine, E.; Leprince, J.; Tomé, J.-P.; Dessen, P.; Vaudry, H.; Misrahi, M. The basolateral sorting signals of the thyrotropin and luteinizing hormone receptors: An unusual family of signals sharing an unusual distal intracellular localization, but unrelated in their structures. *Mol. Endocrinol.* **2004**, *18*, 733–746. [CrossRef]

51. Nadler, L.S.; Kumar, G.; Nathanson, N.M. Identification of a basolateral sorting signal for the M3 muscarinic acetylcholine receptor in Madin-Darby canine kidney cells. *J. Biol. Chem.* **2001**, *276*, 10539–10547. [CrossRef]

52. Iverson, H.A.; Fox, D., 3rd; Nadler, L.S.; Klevit, R.E.; Nathanson, N.M. Identification and structural determination of the M(3) muscarinic acetylcholine receptor basolateral sorting signal. *J. Biol. Chem.* **2005**, *280*, 24568–24575. [CrossRef]

53. King, B.R.; Guda, C. ngLOC: An n-gram-based Bayesian method for estimating the subcellular proteomes of eukaryotes. *Genome Biol.* **2007**, *8*, R68. [CrossRef] [PubMed]

54. Blobel, G. Intracellular protein topogenesis. *Proc. Natl. Acad. Sci. USA* **1980**, *77*, 1496–1500. [CrossRef] [PubMed]

55. Balogh, I.; Maráz, A. Presence of STA gene sequences in brewer's yeast genome. *Lett. Appl. Microbiol.* **2018**, *22*, 400–404. [CrossRef] [PubMed]

56. Lamriben, L.; Graham, J.B.; Adams, B.M.; Hebert, D.N. N-Glycan-based ER molecular chaperone and protein quality control system: The calnexin binding cycle. *Traffic* **2016**, *17*, 308–326. [CrossRef]

57. Saraogi, I.; Shan, S. Molecular mechanism of co-translational protein targeting by the signal recognition particle. *Traffic* **2011**, *12*, 535–542. [CrossRef]

58. Reindl, M.; Hänsch, S.; Weidtkamp-Peters, S.; Schipper, K. A Potential lock-type mechanism for unconventional secretion in fungi. *Int. J. Mol. Sci.* **2019**, *20*, 460. [CrossRef]

59. Chartron, J.W.; Gonzalez, G.M.; Clemons Jr, W.M. A structural model of the Sgt2 protein and its interactions with chaperones and the Get4/Get5 complex. *J. Biol. Chem.* **2011**, *286*, 34325–34334. [CrossRef]

60. Grudnik, P.; Bange, G.; Sinning, I. Protein targeting by the signal recognition particle. *Biol. Chem.* **2009**, *390*, 775–782. [CrossRef]

61. Keenan, R.J.; Freymann, D.M.; Stroud, R.M.; Walter, P. The signal recognition particle. *Annu. Rev. Biochem.* **2001**, *70*, 755–775. [CrossRef]

62. Borgese, N.; Colombo, S.; Pedrazzini, E. The tale of tail-anchored proteins: Coming from the cytosol and looking for a membrane. *J. Cell Biol.* **2003**, *161*, 1013–1019. [CrossRef]

63. Hartl, F.U.; Hayer-Hartl, M. Converging concepts of protein folding in vitro and in vivo. *Nat. Struct. Mol. Biol.* **2009**, *16*, 574. [CrossRef]

64. Marzec, M.; Eletto, D.; Argon, Y. GRP94: An HSP90-like protein specialized for protein folding and quality control in the endoplasmic reticulum. *Biochim. Biophys. Acta* **2012**, *1823*, 774–787. [CrossRef] [PubMed]

65. Tannous, A.; Pisoni, G.B.; Hebert, D.N.; Molinari, M. N-linked sugar-regulated protein folding and quality control in the ER. *Semin. Cell Dev. Biol.* **2015**, *41*, 79–89. [CrossRef] [PubMed]

66. Simons, J.F.; Ferro-Novick, S.; Rose, M.D.; Helenius, A. BiP/Kar2p serves as a molecular chaperone during carboxypeptidase Y folding in yeast. *J. Cell Biol.* **1995**, *130*, 41–49. [CrossRef]

67. Hurtley, S.M.; Bole, D.G.; Hoover-Litty, H.; Helenius, A.; Copeland, C.S. Interactions of misfolded influenza virus hemagglutinin with binding protein (BiP). *J. Cell Biol.* **1989**, *108*, 2117–2126. [CrossRef]

68. Machamer, C.E.; Doms, R.W.; Bole, D.G.; Helenius, A.; Rose, J.K. Heavy chain binding protein recognizes incompletely disulfide-bonded forms of vesicular stomatitis virus G protein. *J. Biol. Chem.* **1990**, *265*, 6879–6883.

69. Marquardt, T.; Helenius, A. Misfolding and aggregation of newly synthesized proteins in the endoplasmic reticulum. *J. Cell Biol.* **1992**, *117*, 505–513. [CrossRef]

70. Singh, I.; Doms, R.W.; Wagner, K.R.; Helenius, A. Intracellular transport of soluble and membrane-bound glycoproteins: Folding, assembly and secretion of anchor-free influenza hemagglutinin. *EMBO J.* **1990**, *9*, 631–639. [CrossRef]

71. Hammond, C.; Helenius, A. Quality control in the secretory pathway: Retention of a misfolded viral membrane glycoprotein involves cycling between the ER, intermediate compartment, and golgi apparatus. *J. Cell Biol.* **1994**, *126*, 41–52. [CrossRef]

72. Pincus, D.; Chevalier, M.W.; Aragón, T.; van Anken, E.; Vidal, S.E.; El-Samad, H.; Walter, P. BiP binding to the ER-stress sensor Ire1 tunes the homeostatic behavior of the unfolded protein response. *PLoS Biol.* **2010**, *8*, e1000415. [CrossRef]

73. Schönbrunner, E.R.; Schmid, F.X. Peptidyl-prolyl cis-trans isomerase improves the efficiency of protein disulfide isomerase as a catalyst of protein folding. *Proc. Natl. Acad. Sci. USA* **1992**, *89*, 4510–4513. [CrossRef] [PubMed]

74. Cheng, H.N.; Bovey, F.A. Cis-Trans equilibrium and kinetic studies of acetyl-L-proline and glycyl-L-proline. *Biopolymers* **2018**, *16*, 1465–1472. [CrossRef] [PubMed]

75. Horibe, T.; Yosho, C.; Okada, S.; Tsukamoto, M.; Nagai, H.; Hagiwara, Y.; Tujimoto, Y.; Kikuchi, M. The chaperone activity of protein disulfide isomerase is affected by cyclophilin B and cyclosporin a *In vitro*. *J. Biochem.* **2002**, *132*, 401–407. [CrossRef] [PubMed]

76. Curran, J.; Mohler, P.J. Alternative paradigms for ion channelopathies: Disorders of ion channel membrane trafficking and posttranslational modification. *Annu. Rev. Physiol.* **2015**, *77*, 505–524. [CrossRef]

77. Shao, S.; Hegde, R.S. Membrane protein insertion at the endoplasmic reticulum. *Annu. Rev. Cell Dev. Biol.* **2011**, *27*, 25–56. [CrossRef]

78. Berner, N.; Reutter, K.-R.; Wolf, D.H. Protein quality control of the endoplasmic reticulum and ubiquitin-proteasome-triggered degradation of aberrant proteins: Yeast pioneers the path. *Annu. Rev. Biochem.* **2018**, *87*, 751–782. [CrossRef]

79. Powers, E.T.; Morimoto, R.I.; Dillin, A.; Kelly, J.W.; Balch, W.E. Biological and chemical approaches to diseases of proteostasis deficiency. *Annu. Rev. Biochem.* **2009**, *78*, 959–991. [CrossRef]

80. Labbadia, J.; Morimoto, R.I. The biology of proteostasis in aging and disease. *Annu. Rev. Biochem.* **2015**, *84*, 435–464. [CrossRef]

81. Walker, L.C. Proteopathic strains and the heterogeneity of neurodegenerative diseases. *Annu. Rev. Genet.* **2016**, *50*, 329–346. [CrossRef]

82. Aguzzi, A.; O'Connor, T. Protein aggregation diseases: Pathogenicity and therapeutic perspectives. *Nat. Rev. Drug Discov.* **2010**, *9*, 237. [CrossRef]

83. Stefani, M.; Dobson, C.M. Protein aggregation and aggregate toxicity: New insights into protein folding, misfolding diseases and biological evolution. *J. Mol. Med.* **2003**, *81*, 678–699. [CrossRef] [PubMed]

84. Walter, P.; Ron, D. The unfolded protein response: From stress pathway to homeostatic regulation. *Science* **2011**, *334*, 1081–1086. [CrossRef] [PubMed]

85. Bravo, R.; Parra, V.; Gatica, D.; Rodriguez, A.E.; Torrealba, N.; Paredes, F.; Wang, Z.V.; Zorzano, A.; Hill, J.A.; Jaimovich, E.; et al. Endoplasmic reticulum and the unfolded protein response: Dynamics and metabolic integration. *Int. Rev. Cell Mol. Biol.* **2013**, *301*, 215–290. [CrossRef] [PubMed]

86. Thibault, G.; Ismail, N.; Ng, D.T.W. The unfolded protein response supports cellular robustness as a broad-spectrum compensatory pathway. *Proc. Natl. Acad. Sci. USA* **2011**, *108*, 20597–20602. [CrossRef]

87. Kopp, M.C.; Larburu, N.; Durairaj, V.; Adams, C.J.; Ali, M.M.U. UPR proteins IRE1 and PERK switch BiP from chaperone to ER stress sensor. *Nat. Struct. Mol. Biol.* **2019**, *26*, 1053–1062. [CrossRef] [PubMed]

88. Nedelsky, N.B.; Todd, P.K.; Taylor, J.P. Autophagy and the ubiquitin-proteasome system: Collaborators in neuroprotection. *Biochim. Biophys. Acta* **2008**, *1782*, 691–699. [CrossRef]

89. Smith, M.H.; Ploegh, H.L.; Weissman, J.S. Road to ruin: Targeting proteins for degradation in the endoplasmic reticulum. *Science* **2011**, *334*, 1086–1090. [CrossRef]

90. Varshavsky, A. The ubiquitin system, an immense realm. *Annu. Rev. Biochem.* **2012**, *81*, 167–176. [CrossRef]

91. Romine, I.C.; Wiseman, R.L. PERK Signaling regulates extracellular proteostasis of an amyloidogenic protein during endoplasmic reticulum stress. *Sci. Rep.* **2019**, *9*, 410. [CrossRef]

92. Lu, P.D.; Harding, H.P.; Ron, D. Translation reinitiation at alternative open reading frames regulates gene expression in an integrated stress response. *J. Cell Biol.* **2004**, *167*, 27–33. [CrossRef]

93. Walter, F.; O'Brien, A.; Concannon, C.G.; Düssmann, H.; Prehn, J.H.M. ER stress signaling has an activating transcription factor 6α (ATF6)-dependent "off-switch". *J. Biol. Chem.* **2018**, *293*, 18270–18284. [CrossRef]

94. Lee, K.; Tirasophon, W.; Shen, X.; Michalak, M.; Prywes, R.; Okada, T.; Yoshida, H.; Mori, K.; Kaufman, R.J. IRE1-mediated unconventional mRNA splicing and S2P-mediated ATF6 cleavage merge to regulate XBP1 in signaling the unfolded protein response. *Genes Dev.* **2002**, *16*, 452–466. [CrossRef] [PubMed]

95. Nishikawa, S.I.; Fewell, S.W.; Kato, Y.; Brodsky, J.L.; Endo, T. Molecular chaperones in the yeast endoplasmic reticulum maintain the solubility of proteins for retrotranslocation and degradation. *J. Cell Biol.* **2001**, *153*, 1061–1070. [CrossRef] [PubMed]

96. Elsasser, S.; Finley, D. Delivery of ubiquitinated substrates to protein-unfolding machines. *Nat. Cell Biol.* **2005**, *7*, 742–749. [CrossRef] [PubMed]

97. Jakob, C.A.; Burda, P.; Roth, J.; Aebi, M. Degradation of misfolded endoplasmic reticulum glycoproteins in Saccharomyces cerevisiae is determined by a specific oligosaccharide structure. *J. Cell Biol.* **1998**, *142*, 1223–1233. [CrossRef] [PubMed]

98. Sato, S.; Ward, C.L.; Krouse, M.E.; Wine, J.J.; Kopito, R.R. Glycerol reverses the misfolding phenotype of the most common cystic fibrosis mutation. *J. Biol. Chem.* **1996**, *271*, 635–638. [CrossRef]

99. Wilkinson, S. ER-phagy: Shaping up and destressing the endoplasmic reticulum. *FEBS J.* **2019**, *286*, 2645–2663. [CrossRef]

100. Jain, B.P. An overview of unfolded protein response signaling and its role in cancer. *Cancer Biother. Radiopharm.* **2017**, *32*, 275–281. [CrossRef]

101. Kaushik, S.; Cuervo, A.M. Chaperone-mediated autophagy: A unique way to enter the lysosome world. *Trends Cell Biol.* **2012**, *22*, 407–417. [CrossRef]

102. Fu, S.; Watkins, S.M.; Hotamisligil, G.S. The role of endoplasmic reticulum in hepatic lipid homeostasis and stress signaling. *Cell Metab.* **2012**, *15*, 623–634. [CrossRef]

103. Garg, A.D.; Kaczmarek, A.; Krysko, O.; Vandenabeele, P.; Krysko, D.V.; Agostinis, P. ER stress-induced inflammation: Does it aid or impede disease progression? *Trends Mol. Med.* **2012**, *18*, 589–598. [CrossRef] [PubMed]

104. Li, H.; Korennykh, A.V.; Behrman, S.L.; Walter, P. Mammalian endoplasmic reticulum stress sensor IRE1 signals by dynamic clustering. *Proc. Natl. Acad. Sci. USA* **2010**, *107*, 16113–16118. [CrossRef] [PubMed]

105. Urano, F.; Wang, X.; Bertolotti, A.; Zhang, Y.; Chung, P.; Harding, H.P.; Ron, D. Coupling of stress in the ER to activation of JNK protein kinases by transmembrane protein kinase IRE1. *Science* **2000**, *287*, 664. [CrossRef] [PubMed]

106. Valencia-Sanchez, M.A.; Liu, J.; Hannon, G.J.; Parker, R. Control of translation and mRNA degradation by miRNAs and siRNAs. *Genes Dev.* **2006**, *20*, 515–524. [CrossRef] [PubMed]

107. Hara, T.; Hashimoto, Y.; Akuzawa, T.; Hirai, R.; Kobayashi, H.; Sato, K. Rer1 and calnexin regulate endoplasmic reticulum retention of a peripheral myelin protein 22 mutant that causes type 1A Charcot-Marie-Tooth disease. *Sci. Rep.* **2014**, *4*, 6992. [CrossRef] [PubMed]

108. Briant, K.; Johnson, N.; Swanton, E. Transmembrane domain quality control systems operate at the endoplasmic reticulum and Golgi apparatus. *PLoS ONE* **2017**, *12*, e0173924. [CrossRef]

109. Okiyoneda, T.; Veit, G.; Sakai, R.; Aki, M.; Fujihara, T.; Higashi, M.; Susuki-Miyata, S.; Miyata, M.; Fukuda, N.; Yoshida, A.; et al. Chaperone-independent peripheral quality control of CFTR by RFFL E3 Ligase. *Dev. Cell* **2018**, *44*, 694–708.e7. [CrossRef]

110. Tzivoni, D.; Gavish, A.; Zin, D.; Gottlieb, S.; Moriel, M.; Keren, A.; Banai, S.; Stern, S. Prognostic significance of ischemic episodes in patients with previous myocardial infarction. *Am. J. Cardiol.* **1988**, *62*, 661–664. [CrossRef]

111. Fath, S.; Mancias, J.D.; Bi, X.; Goldberg, J. Structure and organization of coat proteins in the COPII cage. *Cell* **2007**, *129*, 1325–1336. [CrossRef] [PubMed]

112. Li, J.; Ahat, E.; Wang, Y. Golgi structure and function in health, stress, and diseases. *Results Probl. Cell Differ.* **2019**, *67*, 441–485. [CrossRef]

113. McCaughey, J.; Stephens, D.J. COPII-dependent ER export in animal cells: Adaptation and control for diverse cargo. *Histochem. Cell Biol.* **2018**, *150*, 119–131. [CrossRef] [PubMed]

114. Huotari, J.; Helenius, A. Endosome maturation. *EMBO J.* **2011**, *30*, 3481–3500. [CrossRef] [PubMed]

115. Cocucci, E.; Meldolesi, J. Ectosomes and exosomes: Shedding the confusion between extracellular vesicles. *Trends Cell Biol.* **2015**, *25*, 364–372. [CrossRef] [PubMed]

116. Pegtel, D.M.; Cosmopoulos, K.; Thorley-Lawson, D.A.; van Eijndhoven, M.A.J.; Hopmans, E.S.; Lindenberg, J.L.; de Gruijl, T.D.; Würdinger, T.; Middeldorp, J.M. Functional delivery of viral miRNAs via exosomes. *Proc. Natl. Acad. Sci. USA* **2010**, *107*, 6328–6333. [CrossRef] [PubMed]

117. Valadi, H.; Ekström, K.; Bossios, A.; Sjöstrand, M.; Lee, J.J.; Lötvall, J.O. Exosome-mediated transfer of mRNAs and microRNAs is a novel mechanism of genetic exchange between cells. *Nat. Cell Biol.* **2007**, *9*, 654–659. [CrossRef] [PubMed]

118. Takeuchi, T.; Suzuki, M.; Fujikake, N.; Popiel, H.A.; Kikuchi, H.; Futaki, S.; Wada, K.; Nagai, Y. Intercellular chaperone transmission via exosomes contributes to maintenance of protein homeostasis at the organismal level. *Proc. Natl. Acad. Sci. USA* **2015**, *112*, E2497–E2506. [CrossRef]

119. Li, X.; Corbett, A.L.; Taatizadeh, E.; Tasnim, N.; Little, J.P.; Garnis, C.; Daugaard, M.; Guns, E.; Hoorfar, M.; Li, I.T.S. Challenges and opportunities in exosome research-Perspectives from biology, engineering, and cancer therapy. *APL Bioeng.* **2019**, *3*, 11503. [CrossRef]

120. Ma, D.; Zerangue, N.; Lin, Y.-F.; Collins, A.; Yu, M.; Jan, Y.N.; Jan, L.Y. Role of ER export signals in controlling surface potassium channel numbers. *Science* **2001**, *291*, 316–319. [CrossRef] [PubMed]

121. Petrecca, K.; Atanasiu, R.; Akhavan, A.; Shrier, A. N-linked glycosylation sites determine HERG channel surface membrane expression. *J. Physiol.* **1999**, *515 Pt 1*, 41–48. [CrossRef]

122. Watanabe, I.; Wang, H.-G.; Sutachan, J.J.; Zhu, J.; Recio-Pinto, E.; Thornhill, W.B. Glycosylation affects rat Kv1.1 potassium channel gating by a combined surface potential and cooperative subunit interaction mechanism. *J. Physiol.* **2003**, *550*, 51–66. [CrossRef]

123. Lopez-Rodriguez, A.; Holmgren, M. Restoration of proper trafficking to the cell surface for membrane proteins harboring cysteine mutations. *PLoS ONE* **2012**, *7*, e47693. [CrossRef] [PubMed]

124. Misonou, H.; Mohapatra, D.P.; Park, E.W.; Leung, V.; Zhen, D.; Misonou, K.; Anderson, A.E.; Trimmer, J.S. Regulation of ion channel localization and phosphorylation by neuronal activity. *Nat. Neurosci.* **2004**, *7*, 711–718. [CrossRef] [PubMed]

125. Carneiro, A.M.; Ingram, S.L.; Beaulieu, J.-M.; Sweeney, A.; Amara, S.G.; Thomas, S.M.; Caron, M.G.; Torres, G.E. The multiple LIM domain-containing adaptor protein Hic-5 synaptically colocalizes and interacts with the dopamine transporter. *J. Neurosci.* **2002**, *22*, 7045–7054. [CrossRef] [PubMed]

126. Carneiro, A.M.D.; Blakely, R.D. Serotonin-, protein kinase C-, and Hic-5-associated redistribution of the platelet serotonin transporter. *J. Biol. Chem.* **2006**, *281*, 24769–24780. [CrossRef]

127. Offringa, R.; Huang, F. Phosphorylation-dependent trafficking of plasma membrane proteins in animal and plant cells. *J. Integr. Plant Biol.* **2013**, *55*, 789–808. [CrossRef]

128. Martínez-Mármol, R.; Comes, N.; Styrczewska, K.; Pérez-Verdaguer, M.; Vicente, R.; Pujadas, L.; Soriano, E.; Sorkin, A.; Felipe, A. Unconventional EGF-induced ERK1/2-mediated Kv1.3 endocytosis. *Cell. Mol. Life Sci.* **2016**, *73*, 1515–1528. [CrossRef]

129. Wang, Y.; Yang, J.; Yi, J. Redox sensing by proteins: Oxidative modifications on cysteines and the consequent events. *Antioxid. Redox Signal.* **2012**, *16*, 649–657. [CrossRef]

130. Bindoli, A.; Fukuto, J.M.; Forman, H.J. Thiol chemistry in peroxidase catalysis and redox signaling. *Antioxid. Redox Signal.* **2008**, *10*, 1549–1564. [CrossRef]

131. Bogeski, I.; Niemeyer, B.A. Redox regulation of ion channels. *Antioxid. Redox Signal.* **2014**, *21*, 859–862. [CrossRef]

132. Chen, B.; Sun, Y.; Niu, J.; Jarugumilli, G.K.; Wu, X. Protein lipidation in cell signaling and diseases: Function, regulation, and therapeutic opportunities. *Cell Chem. Biol.* **2018**, *25*, 817–831. [CrossRef]

133. Jeffries, O.; Geiger, N.; Rowe, I.C.M.; Tian, L.; McClafferty, H.; Chen, L.; Bi, D.; Knaus, H.G.; Ruth, P.; Shipston, M.J. Palmitoylation of the S0-S1 linker regulates cell surface expression of voltage- and calcium-activated potassium (BK) channels. *J. Biol. Chem.* **2010**, *285*, 33307–33314. [CrossRef] [PubMed]

134. Tian, L.; McClafferty, H.; Knaus, H.-G.; Ruth, P.; Shipston, M.J. Distinct acyl protein transferases and thioesterases control surface expression of calcium-activated potassium channels. *J. Biol. Chem.* **2012**, *287*, 14718–14725. [CrossRef] [PubMed]

135. Alioua, A.; Li, M.; Wu, Y.; Stefani, E.; Toro, L. Unconventional myristoylation of large-conductance Ca^{2+}-activated K^+ channel (Slo1) via serine/threonine residues regulates channel surface expression. *Proc. Natl. Acad. Sci. USA* **2011**, *108*, 10744–10749. [CrossRef] [PubMed]

136. Mizushima, N.; Komatsu, M. Autophagy: Renovation of cells and tissues. *Cell* **2011**, *147*, 728–741. [CrossRef]

137. Johansen, T.; Lamark, T. Selective autophagy: ATG8 family proteins, LIR motifs and cargo receptors. *J. Mol. Biol.* **2020**, *432*, 80–103. [CrossRef]

138. Popovic, D.; Vucic, D.; Dikic, I. Ubiquitination in disease pathogenesis and treatment. *Nat. Med.* **2014**, *20*, 1242–1253. [CrossRef]

139. Herskowitz, I. Functional inactivation of genes by dominant negative mutations. *Nature* **1987**, *329*, 219–222. [CrossRef]

140. Fornace, A.J., Jr.; Nebert, D.W.; Hollander, M.C.; Luethy, J.D.; Papathanasiou, M.; Fargnoli, J.; Holbrook, N.J. Mammalian genes coordinately regulated by growth arrest signals and DNA-damaging agents. *Mol. Cell. Biol.* **1989**, *9*, 4196–4203. [CrossRef]

141. Schmitt-Ney, M.; Habener, J.F. CHOP/GADD153 gene expression response to cellular stresses inhibited by prior exposure to ultraviolet light wavelength band C (UVC). Inhibitory sequence mediating the UVC response localized to exon 1. *J. Biol. Chem.* **2000**, *275*, 40839–40845. [CrossRef]

142. Spear, E.; Ng, D.T. The unfolded protein response: No longer just a special teams player. *Traffic* **2001**, *2*, 515–523. [CrossRef]

143. Vembar, S.S.; Brodsky, J.L. One step at a time: Endoplasmic reticulum-associated degradation. *Nat. Rev. Mol. Cell Biol.* **2008**, *9*, 944–957. [CrossRef] [PubMed]

144. Marciniak, S.J.; Ron, D. Endoplasmic reticulum stress signaling in disease. *Physiol. Rev.* **2006**, *86*, 1133–1149. [CrossRef] [PubMed]

145. Uehara, T.; Nakamura, T.; Yao, D.; Shi, Z.-Q.; Gu, Z.; Ma, Y.; Masliah, E.; Nomura, Y.; Lipton, S.A. S-Nitrosylated protein-disulphide isomerase links protein misfolding to neurodegeneration. *Nature* **2006**, *441*, 513. [CrossRef] [PubMed]

146. Zhang, K.; Kaufman, R.J. The unfolded protein response. A stress signaling pathway critical for health and disease. *Neurology* **2006**, *66*, S102–S109. [CrossRef]

147. Otsu, M.; Sitia, R. Diseases originating from altered protein quality control in the endoplasmic reticulum. *Curr. Med. Chem.* **2007**, *14*, 1639–1652. [CrossRef]

148. Valastyan, J.S.; Lindquist, S. Mechanisms of protein-folding diseases at a glance. *Dis. Model. Mech.* **2014**, *7*, 9–14. [CrossRef]

149. Foufelle, F.; Ferré, P. Unfolded protein response: Its role in physiology and physiopathology TT - La réponse UPR: Son rôle physiologique et physiopathologique. *Med. Sci.* **2007**, *23*, 291–296. [CrossRef]

150. Menzies, F.M.; Moreau, K.; Rubinsztein, D.C. Protein misfolding disorders and macroautophagy. *Curr. Opin. Cell Biol.* **2011**, *23*, 190–197. [CrossRef]

151. Wang, S.; Kaufman, R.J. The impact of the unfolded protein response on human disease. *J. Cell Biol.* **2012**, *197*, 857–867. [CrossRef]

152. Anfinsen, C.B. Principles that govern the folding of protein chains. *Science* **1973**, *181*, 223–230. [CrossRef]

153. Lee, C.; Ham, S. Characterizing amyloid-beta protein misfolding from molecular dynamics simulations with explicit water. *J. Comput. Chem.* **2011**, *32*, 349–355. [CrossRef]

154. Huber, R.; Carrell, R.W. Implications of the three-dimensional structure of alpha 1-antitrypsin for structure and function of serpins. *Biochemistry* **1989**, *28*, 8951–8966. [CrossRef] [PubMed]

155. Carr, C.M.; Chaudhry, C.; Kim, P.S. Influenza hemagglutinin is spring-loaded by a metastable native conformation. *Proc. Natl. Acad. Sci. USA* **1997**, *94*, 14306–14313. [CrossRef] [PubMed]

156. Bullough, P.A.; Hughson, F.M.; Skehel, J.J.; Wiley, D.C. Structure of influenza haemagglutinin at the pH of membrane fusion. *Nature* **1994**, *371*, 37–43. [CrossRef]

157. Orosz, A.; Wisniewski, J.; Wu, C. Regulation of Drosophila heat shock factor trimerization: Global sequence requirements and independence of nuclear localization. *Mol. Cell. Biol.* **1996**, *16*, 7018–7030. [CrossRef] [PubMed]

158. Hipp, M.S.; Park, S.-H.; Hartl, F.U. Proteostasis impairment in protein-misfolding and -aggregation diseases. *Trends Cell Biol.* **2014**, *24*, 506–514. [CrossRef]

159. Siddiqui, M.H.; Al-Khaishany, M.Y.; Al-Qutami, M.A.; Al-Whaibi, M.H.; Grover, A.; Ali, H.M.; Al-Wahibi, M.S. Morphological and physiological characterization of different genotypes of faba bean under heat stress. *Saudi J. Biol. Sci.* **2015**, *22*, 656–663. [CrossRef]

160. Currais, A.; Fischer, W.; Maher, P.; Schubert, D. Intraneuronal protein aggregation as a trigger for inflammation and neurodegeneration in the aging brain. *FASEB J.* **2017**, *31*, 5–10. [CrossRef]

161. Tuite, M.F.; Melki, R. Protein misfolding and aggregation in ageing and disease: Molecular processes and therapeutic perspectives. *Prion* **2007**, *1*, 116–120. [CrossRef]

162. Dowhan, W. Molecular basis for membrane phospholipid diversity: Why are there so many lipids? *Annu. Rev. Biochem.* **1997**, *66*, 199–232. [CrossRef]

163. Sanders, C.R.; Nagy, J.K. Misfolding of membrane proteins in health and disease: The lady or the tiger? *Curr. Opin. Struct. Biol.* **2000**, *10*, 438–442. [CrossRef]

164. Sarnataro, D.; Campana, V.; Paladino, S.; Stornaiuolo, M.; Nitsch, L.; Zurzolo, C. PrP(C) association with lipid rafts in the early secretory pathway stabilizes its cellular conformation. *Mol. Biol. Cell* **2004**, *15*, 4031–4042. [CrossRef]

165. Campana, V.; Sarnataro, D.; Fasano, C.; Casanova, P.; Paladino, S.; Zurzolo, C. Detergent-resistant membrane domains but not the proteasome are involved in the misfolding of a PrP mutant retained in the endoplasmic reticulum. *J. Cell Sci.* **2006**, *119*, 433–442. [CrossRef]

166. Wieland, F.T.; Gleason, M.L.; Serafini, T.A.; Rothman, J.E. The rate of bulk flow from the endoplasmic reticulum to the cell surface. *Cell* **1987**, *50*, 289–300. [CrossRef]

167. Mizuno, M.; Singer, S.J. A soluble secretory protein is first concentrated in the endoplasmic reticulum before transfer to the Golgi apparatus. *Proc. Natl. Acad. Sci. USA* **1993**, *90*, 5732–5736. [CrossRef]

168. Nishimura, N.; Balch, W.E. A di-acidic signal required for selective export from the endoplasmic reticulum. *Science* **1997**, *277*, 556–558. [CrossRef] [PubMed]

169. Martínez-Menárguez, J.A.; Geuze, H.J.; Slot, J.W.; Klumperman, J. Vesicular tubular clusters between the ER and Golgi mediate concentration of soluble secretory proteins by exclusion from COPI-coated vesicles. *Cell* **1999**, *98*, 81–90. [CrossRef]

170. Bowie, J.U. Solving the membrane protein folding problem. *Nature* **2005**, *438*, 581–589. [CrossRef]

171. Powl, A.M.; East, J.M.; Lee, A.G. Lipid–protein interactions studied by introduction of a tryptophan residue: the mechanosensitive channel MscL. *Biochemistry* **2003**, *42*, 14306–14317. [CrossRef]

172. Hong, H. Role of lipids in folding, misfolding and function of integral membrane proteins. *Adv. Exp. Med. Biol.* **2015**, *855*, 1–31. [CrossRef]

173. Soto, C.; Pritzkow, S. Protein misfolding, aggregation, and conformational strains in neurodegenerative diseases. *Nat. Neurosci.* **2018**, *21*, 1332–1340. [CrossRef] [PubMed]

174. Marsh, J.A.; Hernández, H.; Hall, Z.; Ahnert, S.E.; Perica, T.; Robinson, C.V.; Teichmann, S.A. Protein complexes are under evolutionary selection to assemble via ordered pathways. *Cell* **2013**, *153*, 461–470. [CrossRef] [PubMed]

175. Ng, D.P.; Poulsen, B.E.; Deber, C.M. Membrane protein misassembly in disease. *Biochim. Biophys. Acta* **2012**, *1818*, 1115–1122. [CrossRef] [PubMed]

176. Braverman, N.E.; D'Agostino, M.D.; Maclean, G.E. Peroxisome biogenesis disorders: Biological, clinical and pathophysiological perspectives. *Dev. Disabil. Res. Rev.* **2013**, *17*, 187–196. [CrossRef]

177. Farré, J.-C.; Mahalingam, S.S.; Proietto, M.; Subramani, S. Peroxisome biogenesis, membrane contact sites, and quality control. *EMBO Rep.* **2019**, *20*, e46864. [CrossRef]

178. Chu, C.Y.; King, J.; Berrini, M.; Rumley, A.C.; Apaja, P.M.; Lukacs, G.L.; Alexander, R.T.; Cordat, E. Degradation mechanism of a Golgi-retained distal renal tubular acidosis mutant of the kidney anion exchanger 1 in renal cells. *Am. J. Physiol. Cell Physiol.* **2014**, *307*, C296–C307. [CrossRef]

179. Cordat, E.; Kittanakom, S.; Yenchitsomanus, P.-T.; Li, J.; Du, K.; Lukacs, G.L.; Reithmeier, R.A.F. Dominant and recessive distal renal tubular acidosis mutations of kidney anion exchanger 1 induce distinct trafficking defects in MDCK cells. *Traffic* **2006**, *7*, 117–128. [CrossRef]

180. Kopito, R.R.; Sitia, R. Aggresomes and Russell bodies. Symptoms of cellular indigestion? *EMBO Rep.* **2000**, *1*, 225–231. [CrossRef]

181. Ruan, L.; Zhou, C.; Jin, E.; Kucharavy, A.; Zhang, Y.; Wen, Z.; Florens, L.; Li, R. Cytosolic proteostasis through importing of misfolded proteins into mitochondria. *Nature* **2017**, *543*, 443–446. [CrossRef]

182. Leidenheimer, N.J. Cognate ligand chaperoning: A novel mechanism for the post-translational regulation of neurotransmitter receptor biogenesis. *Front. Cell. Neurosci.* **2017**, *11*, 245. [CrossRef]

183. Sipe, J.D.; Benson, M.D.; Buxbaum, J.N.; Ikeda, S.-I.; Merlini, G.; Saraiva, M.J.M.; Westermark, P. Amyloid fibril proteins and amyloidosis: Chemical identification and clinical classification International Society of Amyloidosis 2016 Nomenclature Guidelines. *Amyloid* **2016**, *23*, 209–213. [CrossRef] [PubMed]

184. de Coninck, D.; Schmidt, T.H.; Schloetel, J.-G.; Lang, T. Packing density of the amyloid precursor protein in the cell membrane. *Biophys. J.* **2018**, *114*, 1128–1141. [CrossRef] [PubMed]

185. Fabiani, C.; Antollini, S.S. Alzheimer's disease as a membrane disorder: Spatial cross-talk among beta-amyloid peptides, nicotinic acetylcholine receptors and lipid rafts. *Front. Cell. Neurosci.* **2019**, *13*, 309. [CrossRef] [PubMed]

186. Kollmer, M.; Meinhardt, K.; Haupt, C.; Liberta, F.; Wulff, M.; Linder, J.; Handl, L.; Heinrich, L.; Loos, C.; Schmidt, M.; et al. Electron tomography reveals the fibril structure and lipid interactions in amyloid deposits. *Proc. Natl. Acad. Sci. USA* **2016**, *113*, 5604–5609. [CrossRef] [PubMed]

187. Glover, J.R.; Lindquist, S. Hsp104, Hsp70, and Hsp40: A novel chaperone system that rescues previously aggregated proteins. *Cell* **1998**, *94*, 73–82. [CrossRef]

188. Mas, G.; Hiller, S. Conformational plasticity of molecular chaperones involved in periplasmic and outer membrane protein folding. *FEMS Microbiol. Lett.* **2018**, *365*, fny121. [CrossRef]

189. Hartl, F.U.; Bracher, A.; Hayer-Hartl, M. Molecular chaperones in protein folding and proteostasis. *Nature* **2011**, *475*, 324. [CrossRef]

190. Hou, Z.-S.; Ulloa-Aguirre, A.; Tao, Y.-X. Pharmacoperone drugs: Targeting misfolded proteins causing lysosomal storage-, ion channels-, and G protein-coupled receptors-associated conformational disorders. *Expert Rev. Clin. Pharmacol.* **2018**, *11*, 611–624. [CrossRef]

191. Laskey, R.A.; Honda, B.M.; Mills, A.D.; Finch, J.T. Nucleosomes are assembled by an acidic protein which binds histones and transfers them to DNA. *Nature* **1978**, *275*, 416. [CrossRef]

192. Wayne, N.; Mishra, P.; Bolon, D.N. Hsp90 and client protein maturation. *Methods Mol. Biol.* **2011**, *787*, 33–44. [CrossRef]

193. Frydman, J.; Nimmesgern, E.; Ohtsuka, K.; Hartl, F.U. Folding of nascent polypeptide chains in a high molecular mass assembly with molecular chaperones. *Nature* **1994**, *370*, 111–117. [CrossRef]

194. Ellis, R.J.; Hemmingsen, S.M. Molecular chaperones: Proteins essential for the biogenesis of some macromolecular structures. *Trends Biochem. Sci.* **1989**, *14*, 339–342. [CrossRef]

195. Jacob, P.; Hirt, H.; Bendahmane, A. The heat-shock protein/chaperone network and multiple stress resistance. *Plant Biotechnol. J.* **2017**, *15*, 405–414. [CrossRef] [PubMed]

196. Cortez, L.; Sim, V. The therapeutic potential of chemical chaperones in protein folding diseases. *Prion* **2014**, *8*, 197–202. [CrossRef] [PubMed]

197. Morello, J.P.; Salahpour, A.; Laperrière, A.; Bernier, V.; Arthus, M.F.; Lonergan, M.; Petäjä-Repo, U.; Angers, S.; Morin, D.; Bichet, D.G.; et al. Pharmacological chaperones rescue cell-surface expression and function of misfolded V2 vasopressin receptor mutants. *J. Clin. Investig.* **2000**, *105*, 887–895. [CrossRef] [PubMed]

198. He, L.; Hiller, S. Frustrated interfaces facilitate dynamic interactions between native client proteins and holdase chaperones. *Chembiochem* **2019**, *20*, 2803–2806. [CrossRef]

199. Akerfelt, M.; Morimoto, R.; Sistonen, L. Heat shock factors: Integrators of cell stress, development and lifespan. *Nat. Rev. Mol. Cell Biol.* **2010**, *11*, 545–555. [CrossRef]

200. Hoffmann, J.H.; Linke, K.; Graf, P.C.F.; Lilie, H.; Jakob, U. Identification of a redox-regulated chaperone network. *EMBO J.* **2004**, *23*, 160–168. [CrossRef]

201. Mattoo, R.U.H.; Goloubinoff, P. Molecular chaperones are nanomachines that catalytically unfold misfolded and alternatively folded proteins. *Cell. Mol. Life Sci.* **2014**, *71*, 3311–3325. [CrossRef]

202. Kaiser, C.M.; Chang, H.C.; Agashe, V.R.; Lakshmipathy, S.K.; Etchells, S.A.; Hayer-Hartl, M.; Hartl, F.U.; Barral, J.M. Real-time observation of trigger factor function on translating ribosomes. *Nature* **2006**, *444*, 455–460. [CrossRef] [PubMed]

203. Dobson, C.M. Principles of protein folding, misfolding and aggregation. *Semin. Cell Dev. Biol.* **2004**, *15*, 3–16. [CrossRef] [PubMed]

204. Taipale, M.; Krykbaeva, I.; Koeva, M.; Kayatekin, C.; Westover, K.D.; Karras, G.I.; Lindquist, S. Quantitative analysis of HSP90-client interactions reveals principles of substrate recognition. *Cell* **2012**, *150*, 987–1001. [CrossRef] [PubMed]

205. Vannimenus, J.; Toulouse, G. Theory of the frustration effect. II. Ising spins on a square lattice. *J. Phys. C Solid State Phys.* **1977**, *10*, L537–L542. [CrossRef]

206. Ferreiro, D.U.; Komives, E.A.; Wolynes, P.G. Frustration, function and folding. *Curr. Opin. Struct. Biol.* **2018**, *48*, 68–73. [CrossRef] [PubMed]

207. He, L.; Sharpe, T.; Mazur, A.; Hiller, S. A molecular mechanism of chaperone-client recognition. *Sci. Adv.* **2016**, *2*, e1601625. [CrossRef]

208. Wälti, M.A.; Libich, D.S.; Clore, G.M. Extensive sampling of the cavity of the GroEL nanomachine by protein substrates probed by paramagnetic relaxation enhancement. *J. Phys. Chem. Lett.* **2018**, *9*, 3368–3371. [CrossRef]

209. Gething, M.-J.; Sambrook, J. Protein folding in the cell. *Nature* **1992**, *357*, 57–59. [CrossRef]

210. Mayer, M.P.; Bukau, B. Hsp70 chaperones: Cellular functions and molecular mechanism. *Cell. Mol. Life Sci.* **2005**, *62*, 670–684. [CrossRef]

211. Radons, J. The human HSP70 family of chaperones: Where do we stand? *Cell Stress Chaperones* **2016**, *21*, 379–404. [CrossRef]

212. Kampinga, H.H.; Craig, E.A. The HSP70 chaperone machinery: J proteins as drivers of functional specificity. *Nat. Rev. Mol. Cell Biol.* **2010**, *11*, 579–592. [CrossRef]

213. Sanguinetti, M.C.; Tristani-Firouzi, M. hERG potassium channels and cardiac arrhythmia. *Nature* **2006**, *440*, 463–469. [CrossRef] [PubMed]

214. Goldberg, A.L. Protein degradation and protection against misfolded or damaged proteins. *Nature* **2003**, *426*, 895–899. [CrossRef] [PubMed]

215. Hirota, Y.; Kurata, Y.; Kato, M.; Notsu, T.; Koshida, S.; Inoue, T.; Kawata, Y.; Miake, J.; Bahrudin, U.; Li, P.; et al. Functional stabilization of Kv1.5 protein by Hsp70 in mammalian cell lines. *Biochem. Biophys. Res. Commun.* **2008**, *372*, 469–474. [CrossRef] [PubMed]

216. Choo-Kang, L.R.; Zeitlin, P.L. Induction of HSP70 promotes DeltaF508 CFTR trafficking. *Am. J. Physiol. Lung Cell. Mol. Physiol.* **2001**, *281*, L58–L68. [CrossRef]

217. Farinha, C.M.; Nogueira, P.; Mendes, F.; Penque, D.; Amaral, M.D. The human DnaJ homologue (Hdj)-1/heat-shock protein (Hsp) 40 co-chaperone is required for the in vivo stabilization of the cystic fibrosis transmembrane conductance regulator by Hsp70. *Biochem. J.* **2002**, *366*, 797–806. [CrossRef]

218. Meacham, G.C.; Lu, Z.; King, S.; Sorscher, E.; Tousson, A.; Cyr, D.M. The Hdj-2/Hsc70 chaperone pair facilitates early steps in CFTR biogenesis. *EMBO J.* **1999**, *18*, 1492–1505. [CrossRef]

219. Vila-Carriles, W.H.; Zhou, Z.-H.; Bubien, J.K.; Fuller, C.M.; Benos, D.J. Participation of the chaperone Hsc70 in the trafficking and functional expression of ASIC2 in glioma cells. *J. Biol. Chem.* **2007**, *282*, 34381–34391. [CrossRef]

220. Kapoor, N.; Lee, W.; Clark, E.; Bartoszewski, R.; McNicholas, C.M.; Latham, C.B.; Bebok, Z.; Parpura, V.; Fuller, C.M.; Palmer, C.A.; et al. Interaction of ASIC1 and ENaC subunits in human glioma cells and rat astrocytes. *Am. J. Physiol. Cell Physiol.* **2011**, *300*, C1246–C1259. [CrossRef]

221. Goldfarb, S.B.; Kashlan, O.B.; Watkins, J.N.; Suaud, L.; Yan, W.; Kleyman, T.R.; Rubenstein, R.C. Differential effects of Hsc70 and Hsp70 on the intracellular trafficking and functional expression of epithelial sodium channels. *Proc. Natl. Acad. Sci. USA* **2006**, *103*, 5817–5822. [CrossRef]

222. Evans, C.G.; Wisén, S.; Gestwicki, J.E. Heat shock proteins 70 and 90 inhibit early stages of amyloid β-(1–42) aggregation in vitro. *J. Biol. Chem.* **2006**, *281*, 33182–33191. [CrossRef]

223. Moloney, T.C.; Hyland, R.; O'Toole, D.; Paucard, A.; Kirik, D.; O'Doherty, A.; Gorman, A.M.; Dowd, E. Heat shock protein 70 reduces α-synuclein-induced predegenerative neuronal dystrophy in the α-synuclein viral gene transfer rat model of Parkinson's disease. *CNS Neurosci. Ther.* **2013**, *20*, 50–58. [CrossRef] [PubMed]

224. Dedmon, M.M.; Christodoulou, J.; Wilson, M.R.; Dobson, C.M. Heat shock protein 70 inhibits α-synuclein fibril formation via preferential binding to prefibrillar species. *J. Biol. Chem.* **2005**, *280*, 14733–14740. [CrossRef] [PubMed]

225. Flower, T.R.; Chesnokova, L.S.; Froelich, C.A.; Dixon, C.; Witt, S.N. Heat shock prevents alpha-synuclein-induced apoptosis in a yeast model of parkinson's disease. *J. Mol. Biol.* **2005**, *351*, 1081–1100. [CrossRef] [PubMed]

226. Prodromou, C. The "active life" of Hsp90 complexes. *Biochim. Biophys. Acta Mol. Cell Res.* **2012**, *1823*, 614–623. [CrossRef]

227. Karagöz, G.E.; Rüdiger, S.G.D. Hsp90 interaction with clients. *Trends Biochem. Sci.* **2015**, *40*, 117–125. [CrossRef]

228. Taipale, M.; Jarosz, D.F.; Lindquist, S. Hsp90 at the hub of protein homeostasis: Emerging mechanistic insights. *Nat. Rev. Mol. Cell Biol.* **2010**, *11*, 515–528. [CrossRef]

229. Boulon, S.; Bertrand, E.; Pradet-Balade, B. Hsp90 and the R2TP co-chaperone complex: Building multi-protein machineries essential for cell growth and gene expression. *RNA Biol.* **2012**, *9*, 148–155. [CrossRef] [PubMed]

230. Kadota, Y.; Shirasu, K.; Guerois, R. NLR sensors meet at the SGT1-HSP90 crossroad. *Trends Biochem. Sci.* **2010**, *35*, 199–207. [CrossRef]

231. Pearl, L.H.; Prodromou, C.; Workman, P. The Hsp90 molecular chaperone: An open and shut case for treatment. *Biochem. J.* **2008**, *410*, 439–453. [CrossRef]

232. Wegele, H.; Müller, L.; Buchner, J. *Hsp70 and Hsp90—A Relay Team for Protein Folding BT- Reviews of Physiology, Biochemistry and Pharmacology*; Springer: Berlin/Heidelberg, Germany, 2004; pp. 1–44, ISBN 978-3-540-44423-7.

233. Jakob, U.; Lilie, H.; Meyer, I.; Buchner, J. Transient interaction of Hsp90 with early unfolding intermediates of citrate synthase: Implications for heat shock in vivo. *J. Biol. Chem.* **1995**, *270*, 7288–7294. [CrossRef] [PubMed]

234. McLaughlin, S.H.; Smith, H.W.; Jackson, S.E. Stimulation of the weak ATPase activity of human Hsp90 by a client protein. *J. Mol. Biol.* **2002**, *315*, 787–798. [CrossRef] [PubMed]

235. Krukenberg, K.A.; Street, T.O.; Lavery, L.A.; Agard, D.A. Conformational dynamics of the molecular chaperone Hsp90. *Q. Rev. Biophys.* **2011**, *44*, 229–255. [CrossRef] [PubMed]

236. González-del Pozo, M.; Borrego, S.; Barragán, I.; Pieras, J.I.; Santoyo, J.; Matamala, N.; Naranjo, B.; Dopazo, J.; Antiñolo, G. Mutation screening of multiple genes in Spanish patients with autosomal recessive retinitis pigmentosa by targeted resequencing. *PLoS ONE* **2011**, *6*. [CrossRef] [PubMed]

237. Young, J.C.; Agashe, V.R.; Siegers, K.; Hartl, F.U. Pathways of chaperone-mediated protein folding in the cytosol. *Nat. Rev. Mol. Cell Biol.* **2004**, *5*, 781–791. [CrossRef]

238. Takayama, S.; Reed, J.C.; Homma, S. Heat-shock proteins as regulators of apoptosis. *Oncogene* **2003**, *22*, 9041–9047. [CrossRef]

239. Wiech, H.; Buchner, J.; Zimmermann, R.; Jakob, U. Hsp90 chaperones protein folding in vitro. *Nature* **1992**, *358*, 169–170. [CrossRef]

240. Ficker, E.; Dennis, A.T.; Wang, L.; Brown, A.M. Role of the cytosolic chaperones Hsp70 and Hsp90 in maturation of the cardiac potassium channel HERG. *Circ. Res.* **2003**, *92*, e87–e100. [CrossRef]

241. Gao, Y.; Yechikov, S.; Vazquez, A.E.; Chen, D.; Nie, L. Distinct roles of molecular chaperones HSP90α and HSP90β in the biogenesis of KCNQ4 channels. *PLoS ONE* **2013**, *8*, e57282. [CrossRef]

242. Chen, X.; Liu, P.; Wang, Q.; Li, Y.; Fu, L.; Fu, H.; Zhu, J.; Chen, Z.; Zhu, W.; Xie, C.; et al. DCZ3112, a novel Hsp90 inhibitor, exerts potent antitumor activity against HER2-positive breast cancer through disruption of Hsp90-Cdc37 interaction. *Cancer Lett.* **2018**, *434*, 70–80. [CrossRef]

243. Shelton, L.B.; Koren 3rd, J.; Blair, L.J. Imbalances in the Hsp90 Chaperone Machinery: Implications for Tauopathies. *Front. Neurosci.* **2017**, *11*, 724. [CrossRef] [PubMed]

244. Kopito, R.R. Aggresomes, inclusion bodies and protein aggregation. *Trends Cell Biol.* **2000**, *10*, 524–530. [CrossRef]

245. Garcia-Mata, R.; Gao, Y.-S.; Sztul, E. Hassles with taking out the garbage: Aggravating aggresomes. *Traffic* **2002**, *3*, 388–396. [CrossRef] [PubMed]

246. Wilkinson, B.; Gilbert, H.F. Protein disulfide isomerase. *Biochim. Biophys. Acta* **2004**, *1699*, 35–44. [CrossRef]

247. Brehme, M.; Voisine, C.; Rolland, T.; Wachi, S.; Soper, J.H.; Zhu, Y.; Orton, K.; Villella, A.; Garza, D.; Vidal, M.; et al. A chaperome subnetwork safeguards proteostasis in aging and neurodegenerative disease. *Cell Rep.* **2014**, *9*, 1135–1150. [CrossRef]

248. Gruber, C.W.; Cemazar, M.; Heras, B.; Martin, J.L.; Craik, D.J. Protein disulfide isomerase: The structure of oxidative folding. *Trends Biochem. Sci.* **2006**, *31*, 455–464. [CrossRef]

249. Galligan, J.J.; Petersen, D.R. The human protein disulfide isomerase gene family. *Hum. Genomics* **2012**, *6*, 6. [CrossRef]

250. Guna, A.; Hegde, R.S. Transmembrane domain recognition during membrane protein biogenesis and quality control. *Curr. Biol.* **2018**, *28*, R498–R511. [CrossRef]

251. Heyden, M.; Freites, J.A.; Ulmschneider, M.B.; White, S.H.; Tobias, D.J. Assembly and stability of α-helical membrane proteins. *Soft Matter* **2012**, *8*, 7742–7752. [CrossRef]

252. Cymer, F.; von Heijne, G.; White, S.H. Mechanisms of integral membrane protein insertion and folding. *J. Mol. Biol.* **2015**, *427*, 999–1022. [CrossRef]

253. Béguin, P.; Hasler, U.; Beggah, A.; Horisberger, J.D.; Geering, K. Membrane integration of Na,K-ATPase alpha-subunits and beta-subunit assembly. *J. Biol. Chem.* **1998**, *273*, 24921–24931. [CrossRef]

254. Béguin, P.; Hasler, U.; Staub, O.; Geering, K. Endoplasmic reticulum quality control of oligomeric membrane proteins: Topogenic determinants involved in the degradation of the unassembled Na,K-ATPase alpha subunit and in its stabilization by beta subunit assembly. *Mol. Biol. Cell* **2000**, *11*, 1657–1672. [CrossRef] [PubMed]

255. Feige, M.J.; Hendershot, L.M. Quality control of integral membrane proteins by assembly-dependent membrane integration. *Mol. Cell* **2013**, *51*, 297–309. [CrossRef]

256. Lamb, J.R.; Tugendreich, S.; Hieter, P. Tetratrico peptide repeat interactions: To TPR or not to TPR? *Trends Biochem. Sci.* **1995**, *20*, 257–259. [CrossRef]

257. McDonough, H.; Patterson, C. CHIP: A link between the chaperone and proteasome systems. *Cell Stress Chaperones* **2003**, *8*, 303–308. [CrossRef]

258. Apaja, P.M.; Xu, H.; Lukacs, G.L. Quality control for unfolded proteins at the plasma membrane. *J. Cell Biol.* **2010**, *191*, 553–570. [CrossRef] [PubMed]

259. Li, P.; Kurata, Y.; Maharani, N.; Mahati, E.; Higaki, K.; Hasegawa, A.; Shirayoshi, Y.; Yoshida, A.; Kondo, T.; Kurozawa, Y.; et al. E3 ligase CHIP and Hsc70 regulate Kv1.5 protein expression and function in mammalian cells. *J. Mol. Cell. Cardiol.* **2015**, *86*, 138–146. [CrossRef] [PubMed]

260. Denning, G.M.; Anderson, M.P.; Amara, J.F.; Marshall, J.; Smith, A.E.; Welsh, M.J. Processing of mutant cystic fibrosis transmembrane conductance regulator is temperature-sensitive. *Nature* **1992**, *358*, 761. [CrossRef] [PubMed]

261. Lampel, A.; Bram, Y.; Levy-Sakin, M.; Bacharach, E.; Gazit, E. The Effect of chemical chaperones on the assembly and stability of HIV-1 capsid protein. *PLoS ONE* **2013**, *8*, 25–29. [CrossRef]

262. Dandage, R.; Bandyopadhyay, A.; Jayaraj, G.G.; Saxena, K.; Dalal, V.; Das, A.; Chakraborty, K. Classification of chemical chaperones based on their effect on protein folding landscapes. *ACS Chem. Biol.* **2015**, *10*, 813–820. [CrossRef]

263. Perlmutter, D.H. Chemical chaperones: A pharmacological strategy for disorders of protein folding and trafficking. *Pediatr. Res.* **2002**, *52*, 832–836. [CrossRef]

264. Yancey, P.; Clark, M.; Hand, S.; Bowlus, R.; Somero, G. Living with water stress: Evolution of osmolyte systems. *Science* **1982**, *217*, 1214–1222. [CrossRef] [PubMed]

265. Wlodarczyk, S.R.; Custódio, D.; Pessoa Jr, A.; Monteiro, G. Influence and effect of osmolytes in biopharmaceutical formulations. *Eur. J. Pharm. Biopharm.* **2018**, *131*, 92–98. [CrossRef] [PubMed]

266. Lin, T.Y.; Timasheff, S.N. Why do some organisms use a urea-methylamine mixture as osmolyte? Thermodynamic compensation of urea and trimethylamine N-oxide interactions with protein. *Biochemistry* **1994**, *33*, 12695–12701. [CrossRef] [PubMed]

267. Baskakov, I.; Bolen, D.W. Forcing thermodynamically unfolded proteins to fold. *J. Biol. Chem.* **1998**, *273*, 4831–4834. [CrossRef] [PubMed]

268. Street, T.O.; Bolen, D.W.; Rose, G.D. A molecular mechanism for osmolyte-induced protein stability. *Proc. Natl. Acad. Sci. USA* **2006**, *103*, 13997–14002. [CrossRef] [PubMed]

269. Fischer, H.; Fukuda, N.; Barbry, P.; Illek, B.; Sartori, C.; Matthay, M.A. Partial restoration of defective chloride conductance in DeltaF508 CF mice by trimethylamine oxide. *Am. J. Physiol. Lung Cell. Mol. Physiol.* **2001**, *281*, L52–L57. [CrossRef] [PubMed]

270. Deen, P.M.; Marr, N.; Kamsteeg, E.J.; van Balkom, B.W. Nephrogenic diabetes insipidus. *Curr. Opin. Nephrol. Hypertens.* **2000**, *9*, 591–595. [CrossRef]

271. Langley, J.M.; Balfe, J.W.; Selander, T.; Ray, P.N.; Clarke, J.T. Autosomal recessive inheritance of vasopressin-resistant diabetes insipidus. *Am. J. Med. Genet.* **1991**, *38*, 90–94. [CrossRef]

272. Tamarappoo, B.K.; Yang, B.; Verkman, A.S. Misfolding of mutant aquaporin-2 water channels in nephrogenic siabetes insipidus. *J. Biol. Chem.* **1999**, *274*, 34825–34831. [CrossRef]

273. Tamarappoo, B.K.; Verkman, A.S. Defective aquaporin-2 trafficking in nephrogenic diabetes insipidus and correction by chemical chaperones. *J. Clin. Investig.* **1998**, *101*, 2257–2267. [CrossRef]

274. Hayashi, H.; Sugiyama, Y. 4-phenylbutyrate enhances the cell surface expression and the transport capacity of wild-type and mutated bile salt export pumps. *Hepatology* **2007**, *45*, 1506–1516. [CrossRef] [PubMed]

275. Malatack, J.J.; Doyle, D. A Drug regimen for progressive familial cholestasis Type 2. *Pediatrics* **2018**, *141*, e20163877. [CrossRef]

276. Rubenstein, R.C.; Egan, M.E.; Zeitlin, P.L. In vitro pharmacologic restoration of CFTR-mediated chloride transport with sodium 4-phenylbutyrate in cystic fibrosis epithelial cells containing delta F508-CFTR. *J. Clin. Investig.* **1997**, *100*, 2457–2465. [CrossRef]

277. Rubenstein, R.C.; Zeitlin, P.L. A pilot clinical trial of oral sodium 4-phenylbutyrate (Buphenyl) in deltaF508-homozygous cystic fibrosis patients: Partial restoration of nasal epithelial CFTR function. *Am. J. Respir. Crit. Care Med.* **1998**, *157*, 484–490. [CrossRef]

278. Duricka, D.L.; Brown, R.L.; Varnum, M.D. Defective trafficking of cone photoreceptor CNG channels induces the unfolded protein response and ER-stress-associated cell death. *Biochem. J.* **2012**, *441*, 685–696. [CrossRef] [PubMed]

279. Ulloa-Aguirre, A.; Zarinan, T.; Conn, P.M. Pharmacoperones: Targeting therapeutics toward diseases caused by protein misfolding. *Rev. Investig. Clin.* **2015**, *67*, 15–19. [PubMed]

280. Bernier, V.; Lagacé, M.; Bichet, D.G.; Bouvier, M. Pharmacological chaperones: Potential treatment for conformational diseases. *Trends Endocrinol. Metab.* **2004**, *15*, 222–228. [CrossRef] [PubMed]

281. Janovick, J.A.; Goulet, M.; Bush, E.; Greer, J.; Wettlaufer, D.G.; Conn, P.M. Structure-activity relations of successful pharmacologic chaperones for rescue of naturally occurring and manufactured mutants of the gonadotropin-releasing hormone receptor. *J. Pharmacol. Exp. Ther.* **2003**, *305*, 608–614. [CrossRef]

282. Hawtin, S.R. Pharmacological chaperone activity of SR49059 to functionally recover misfolded mutations of the vasopressin V1a receptor. *J. Biol. Chem.* **2006**, *281*, 14604–14614. [CrossRef]

283. Fan, J.-Q. A contradictory treatment for lysosomal storage disorders: Inhibitors enhance mutant enzyme activity. *Trends Pharmacol. Sci.* **2003**, *24*, 355–360. [CrossRef]

284. Ishii, S. Pharmacological chaperone therapy for Fabry disease. *Proc. Jpn. Acad. Ser. B. Phys. Biol. Sci.* **2012**, *88*, 18–30. [CrossRef] [PubMed]

285. Yan, F.; Lin, C.-W.; Weisiger, E.; Cartier, E.A.; Taschenberger, G.; Shyng, S.-L. Sulfonylureas correct trafficking defects of ATP-sensitive potassium channels caused by mutations in the sulfonylurea receptor. *J. Biol. Chem.* **2004**, *279*, 11096–11105. [CrossRef] [PubMed]

286. Bernier, V.; Lagacé, M.; Lonergan, M.; Arthus, M.-F.; Bichet, D.G.; Bouvier, M. Functional rescue of the constitutively internalized V2 vasopressin receptor mutant R137H by the pharmacological chaperone action of SR49059. *Mol. Endocrinol.* **2004**, *18*, 2074–2084. [CrossRef] [PubMed]

287. Bernier, V.; Morello, J.-P.; Zarruk, A.; Debrand, N.; Salahpour, A.; Lonergan, M.; Arthus, M.-F.; Laperrière, A.; Brouard, R.; Bouvier, M.; et al. Pharmacologic chaperones as a potential treatment for X-linked nephrogenic diabetes insipidus. *J. Am. Soc. Nephrol.* **2006**, *17*, 232–243. [CrossRef] [PubMed]

288. Wang, X.-H.; Wang, H.-M.; Zhao, B.-L.; Yu, P.; Fan, Z.-C. Rescue of defective MC4R cell-surface expression and signaling by a novel pharmacoperone Ipsen 17. *J. Mol. Endocrinol.* **2014**, *53*, 17–29. [CrossRef] [PubMed]

289. Smith, J.L.; Reloj, A.R.; Nataraj, P.S.; Bartos, D.C.; Schroder, E.A.; Moss, A.J.; Ohno, S.; Horie, M.; Anderson, C.L.; January, C.T.; et al. Pharmacological correction of long QT-linked mutations in KCNH2 (hERG) increases the trafficking of Kv11.1 channels stored in the transitional endoplasmic reticulum. *Am. J. Physiol. Cell Physiol.* **2013**, *305*, C919–C930. [CrossRef] [PubMed]

290. Delisle, B.P.; Anderson, C.L.; Balijepalli, R.C.; Anson, B.D.; Kamp, T.J.; January, C.T. Thapsigargin selectively rescues the trafficking defective LQT2 channels G601S and F805C. *J. Biol. Chem.* **2003**, *278*, 35749–35754. [CrossRef]

291. Ficker, E.; Obejero-Paz, C.A.; Zhao, S.; Brown, A.M. The binding site for channel blockers that rescue misprocessed human long QT syndrome type 2 ether-a-gogo-related gene (HERG) mutations. *J. Biol. Chem.* **2002**, *277*, 4989–4998. [CrossRef]

292. Los, E.L.; Deen, P.M.T.; Robben, J.H. Potential of nonpeptide (ant)agonists to rescue vasopressin V2 receptor mutants for the treatment of X-linked nephrogenic diabetes insipidus. *J. Neuroendocrinol.* **2010**, *22*, 393–399. [CrossRef]

293. Robben, J.H.; Kortenoeven, M.L.A.; Sze, M.; Yae, C.; Milligan, G.; Oorschot, V.M.; Klumperman, J.; Knoers, N.V.A.M.; Deen, P.M.T. Intracellular activation of vasopressin V2 receptor mutants in nephrogenic diabetes insipidus by nonpeptide agonists. *Proc. Natl. Acad. Sci. USA* **2009**, *106*, 12195–12200. [CrossRef]

294. Jean-Alphonse, F.; Perkovska, S.; Frantz, M.-C.; Durroux, T.; Méjean, C.; Morin, D.; Loison, S.; Bonnet, D.; Hibert, M.; Mouillac, B.; et al. Biased agonist pharmacochaperones of the AVP V2 receptor may treat congenital nephrogenic diabetes insipidus. *J. Am. Soc. Nephrol.* **2009**, *20*, 2190–2203. [CrossRef] [PubMed]

295. White, E.; McKenna, J.; Cavanaugh, A.; Breitwieser, G.E. Pharmacochaperone-mediated rescue of calcium-sensing receptor loss-of-function mutants. *Mol. Endocrinol.* **2009**, *23*, 1115–1123. [CrossRef]

296. Caldwell, R.A.; Grove, D.E.; Houck, S.A.; Cyr, D.M. Increased folding and channel activity of a rare cystic fibrosis mutant with CFTR modulators. *Am. J. Physiol. Lung Cell. Mol. Physiol.* **2011**, *301*, L346–L352. [CrossRef] [PubMed]

297. Deeks, E.D. Lumacaftor/Ivacaftor: A review in cystic fibrosis. *Drugs* **2016**, *76*, 1191–1201. [CrossRef] [PubMed]

298. Liu, Q.; Sabirzhanova, I.; Bergbower, E.A.S.; Yanda, M.; Guggino, W.G.; Cebotaru, L. The CFTR Corrector, VX-809 (Lumacaftor), Rescues ABCA4 trafficking mutants: A potential treatment for Stargardt disease. *Cell. Physiol. Biochem.* **2019**, *53*, 400–412. [CrossRef] [PubMed]

299. Steele, M.; Seiple, W.H.; Carr, R.E.; Klug, R. The clinical utility of visual-evoked potential acuity testing. *Am. J. Ophthalmol.* **1989**, *108*, 572–577. [CrossRef]

300. Ruffin, M.; Roussel, L.; Maillé, É.; Rousseau, S.; Brochiero, E. Vx-809/Vx-770 treatment reduces inflammatory response to Pseudomonas aeruginosa in primary differentiated cystic fibrosis bronchial epithelial cells. *Am. J. Physiol. Lung Cell. Mol. Physiol.* **2018**, *314*, L635–L641. [CrossRef]

301. Chaudary, N. Triplet CFTR modulators: Future prospects for treatment of cystic fibrosis. *Ther. Clin. Risk Manag.* **2018**, *14*, 2375–2383. [CrossRef]

302. Albright, J.D.; Reich, M.F.; Delos Santos, E.G.; Dusza, J.P.; Sum, F.-W.; Venkatesan, A.M.; Coupet, J.; Chan, P.S.; Ru, X.; Mazandarani, H.; et al. 5-Fluoro-2-methyl-N-[4-(5H-pyrrolo[2,1-c]-[1,4]benzodiazepin-10 (11H)-ylcarbonyl)-3- chlorophenyl]benzamide (VPA-985): An orally active arginine vasopressin antagonist with selectivity for V2 receptors. *J. Med. Chem.* **1998**, *41*, 2442–2444. [CrossRef]

303. Robben, J.H.; Sze, M.; Knoers, N.V.A.M.; Deen, P.M.T. Functional rescue of vasopressin V2 receptor mutants in MDCK cells by pharmacochaperones: Relevance to therapy of nephrogenic diabetes insipidus. *Am. J. Physiol. Physiol.* **2007**, *292*, F253–F260. [CrossRef]

304. Moeller, H.B.; Rittig, S.; Fenton, R.A. Nephrogenic diabetes insipidus: Essential insights into the molecular background and potential therapies for treatment. *Endocr. Rev.* **2013**, *34*, 278–301. [CrossRef] [PubMed]

305. Sorrenson, B.; Suetani, R.J.; Williams, M.J.A.; Bickley, V.M.; George, P.M.; Jones, G.T.; McCormick, S.P.A. Functional rescue of mutant ABCA1 proteins by sodium 4-phenylbutyrate. *J. Lipid Res.* **2013**, *54*, 55–62. [CrossRef] [PubMed]

306. Sabirzhanova, I.; Lopes Pacheco, M.; Rapino, D.; Grover, R.; Handa, J.T.; Guggino, W.B.; Cebotaru, L. Rescuing trafficking mutants of the ATP-binding cassette protein, ABCA4, with small molecule correctors as a treatment for Stargardt eye disease. *J. Biol. Chem.* **2015**, *290*, 19743–19755. [CrossRef] [PubMed]

307. Cavanaugh, A.; McKenna, J.; Stepanchick, A.; Breitwieser, G.E. Calcium-sensing receptor biosynthesis includes a cotranslational conformational checkpoint and endoplasmic reticulum retention. *J. Biol. Chem.* **2010**, *285*, 19854–19864. [CrossRef]

308. Dorwart, M.R.; Shcheynikov, N.; Baker, J.M.R.; Forman-Kay, J.D.; Muallem, S.; Thomas, P.J. Congenital chloride-losing diarrhea causing mutations in the STAS domain result in misfolding and mistrafficking of SLC26A3. *J. Biol. Chem.* **2008**, *283*, 8711–8722. [CrossRef]

309. Keller, D.I.; Rougier, J.-S.; Kucera, J.P.; Benammar, N.; Fressart, V.; Guicheney, P.; Madle, A.; Fromer, M.; Schläpfer, J.; Abriel, H. Brugada syndrome and fever: Genetic and molecular characterization of patients carrying SCN5A mutations. *Cardiovasc. Res.* **2005**, *67*, 510–519. [CrossRef]

310. Musil, L.S.; Le, A.-C.N.; VanSlyke, J.K.; Roberts, L.M. Regulation of connexin degradation as a mechanism to increase gap junction assembly and function. *J. Biol. Chem.* **2000**, *275*, 25207–25215. [CrossRef]

311. Vonk, W.I.M.; de Bie, P.; Wichers, C.G.K.; van den Berghe, P.V.E.; van der Plaats, R.; Berger, R.; Wijmenga, C.; Klomp, L.W.J.; van de Sluis, B. The copper-transporting capacity of ATP7A mutants associated with Menkes disease is ameliorated by COMMD1 as a result of improved protein expression. *Cell. Mol. Life Sci.* **2012**, *69*, 149–163. [CrossRef]

312. Nascimento-Ferreira, I.; Santos-Ferreira, T.; Sousa-Ferreira, L.; Auregan, G.; Onofre, I.; Alves, S.; Dufour, N.; Colomer Gould, V.F.; Koeppen, A.; Déglon, N.; et al. Overexpression of the autophagic beclin-1 protein clears mutant ataxin-3 and alleviates Machado-Joseph disease. *Brain* **2011**, *134*, 1400–1415. [CrossRef]

313. Grove, D.E.; Fan, C.-Y.; Ren, H.Y.; Cyr, D.M. The endoplasmic reticulum-associated Hsp40 DNAJB12 and Hsc70 cooperate to facilitate RMA1 E3-dependent degradation of nascent CFTRDeltaF508. *Mol. Biol. Cell* **2011**, *22*, 301–314. [CrossRef]

314. Rowe, S.M.; McColley, S.A.; Rietschel, E.; Li, X.; Bell, S.C.; Konstan, M.W.; Marigowda, G.; Waltz, D.; Boyle, M.P.; Group, V.-. 809-102 S. Lumacaftor/Ivacaftor treatment of patients with cystic fibrosis heterozygous for F508del-CFTR. *Ann. Am. Thorac. Soc.* **2017**, *14*, 213–219. [CrossRef] [PubMed]

315. Gentzsch, M.; Ren, H.Y.; Houck, S.A.; Quinney, N.L.; Cholon, D.M.; Sopha, P.; Chaudhry, I.G.; Das, J.; Dokholyan, N.V.; Randell, S.H.; et al. Restoration of R117H CFTR folding and function in human airway cells through combination treatment with VX-809 and VX-770. *Am. J. Physiol. Cell. Mol. Physiol.* **2016**, *311*, L550–L559. [CrossRef] [PubMed]

316. Bagdany, M.; Veit, G.; Fukuda, R.; Avramescu, R.G.; Okiyoneda, T.; Baaklini, I.; Singh, J.; Sovak, G.; Xu, H.; Apaja, P.M.; et al. Chaperones rescue the energetic landscape of mutant CFTR at single molecule and in cell. *Nat. Commun.* **2017**, *8*, 398. [CrossRef] [PubMed]

317. Wang, Y.; Bartlett, M.C.; Loo, T.W.; Clarke, D.M. Specific rescue of cystic fibrosis transmembrane conductance regulator processing mutants using pharmacological chaperones. *Mol. Pharmacol.* **2006**, *70*, 297–302. [CrossRef] [PubMed]

318. Wellhauser, L.; Chiaw, P.K.; Pasyk, S.; Li, C.; Ramjeesingh, M.; Bear, C.E. A Small-molecule modulator interacts directly with ΔPhe508-CFTR to modify its ATPase activity and conformational stability. *Mol. Pharmacol.* **2009**, *75*, 1430–1438. [CrossRef] [PubMed]

319. Taylor-Cousar, J.L.; Munck, A.; McKone, E.F.; van der Ent, C.K.; Moeller, A.; Simard, C.; Wang, L.T.; Ingenito, E.P.; McKee, C.; Lu, Y.; et al. Tezacaftor–Ivacaftor in patients with cystic fibrosis homozygous for Phe508del. *N. Engl. J. Med.* **2017**, *377*, 2013–2023. [CrossRef]

320. Jacobs, M.T.; Zhang, Y.-W.; Campbell, S.D.; Rudnick, G. Ibogaine, a noncompetitive inhibitor of serotonin transport, acts by stabilizing the cytoplasm-facing State of the transporter. *J. Biol. Chem.* **2007**, *282*, 29441–29447. [CrossRef]

321. Bulling, S.; Schicker, K.; Zhang, Y.-W.; Steinkellner, T.; Stockner, T.; Gruber, C.W.; Boehm, S.; Freissmuth, M.; Rudnick, G.; Sitte, H.H.; et al. The Mechanistic basis for noncompetitive ibogaine inhibition of serotonin and dopamine transporters. *J. Biol. Chem.* **2012**, *287*, 18524–18534. [CrossRef]

322. Sucic, S.; Kasture, A.; Mazhar Asjad, H.M.; Kern, C.; El-Kasaby, A.; Freissmuth, M. When transporters fail to be transported: How to rescue folding-deficient SLC6 transporters. *J. Neurol. Neuromed.* **2016**, *1*, 34–40. [CrossRef]

323. Houck, S.A.; Ren, H.Y.; Madden, V.J.; Bonner, J.N.; Conlin, M.P.; Janovick, J.A.; Conn, P.M.; Cyr, D.M. Quality control autophagy degrades soluble ERAD-resistant conformers of the misfolded membrane protein GnRHR. *Mol. Cell* **2014**, *54*, 166–179. [CrossRef]

324. Leaños-Miranda, A.; Ulloa-Aguirre, A.; Janovick, J.A.; Conn, P.M. In vitro coexpression and pharmacological rescue of mutant gonadotropin-releasing hormone receptors causing hypogonadotropic hypogonadism in humans expressing compound heterozygous alleles. *J. Clin. Endocrinol. Metab.* **2005**, *90*, 3001–3008. [CrossRef] [PubMed]

325. Walker, V.E.; Wong, M.J.H.; Atanasiu, R.; Hantouche, C.; Young, J.C.; Shrier, A. Hsp40 chaperones promote degradation of the HERG potassium channel. *J. Biol. Chem.* **2010**, *285*, 3319–3329. [CrossRef] [PubMed]

326. Zhou, Z.; Gong, Q.; January, C.T. Correction of defective protein trafficking of a mutant HERG potassium channel in human long QT syndrome: Pharmacological and temperature effects. *J. Biol. Chem.* **1999**, *274*, 31123–31126. [CrossRef] [PubMed]

327. Bass, J.; Chiu, G.; Argon, Y.; Steiner, D.F. Folding of insulin receptor monomers is facilitated by the molecular chaperones calnexin and calreticulin and impaired by rapid dimerization. *J. Cell Biol.* **1998**, *141*, 637–646. [CrossRef]

328. Huang, H.; Wang, W.; Tao, Y.-X. Pharmacological chaperones for the misfolded melanocortin-4 receptor associated with human obesity. *Biochim. Biophys. Acta Mol. Basis Dis.* **2017**, *1863*, 2496–2507. [CrossRef] [PubMed]

329. Tao, Y.-X. The melanocortin-4 receptor: Physiology, pharmacology, and pathophysiology. *Endocr. Rev.* **2010**, *31*, 506–543. [CrossRef]

330. Fan, Z.-C.; Tao, Y.-X. Functional characterization and pharmacological rescue of melanocortin-4 receptor mutations identified from obese patients. *J. Cell. Mol. Med.* **2009**, *13*, 3268–3282. [CrossRef]

331. Ulbrich, L.; Favaloro, F.L.; Trobiani, L.; Marchetti, V.; Patel, V.; Pascucci, T.; Comoletti, D.; Marciniak, S.J.; De Jaco, A. Autism-associated R451C mutation in neuroligin3 leads to activation of the unfolded protein response in a PC12 Tet-On inducible system. *Biochem. J.* **2016**, *473*, 423–434. [CrossRef]

332. Shepshelovich, J.; Goldstein-Magal, L.; Globerson, A.; Yen, P.M.; Rotman-Pikielny, P.; Hirschberg, K. Protein synthesis inhibitors and the chemical chaperone TMAO reverse endoplasmic reticulum perturbation induced by overexpression of the iodide transporter pendrin. *J. Cell Sci.* **2005**, *118*, 1577–1586. [CrossRef]

333. Jin, Z.-B.; Mandai, M.; Yokota, T.; Higuchi, K.; Ohmori, K.; Ohtsuki, F.; Takakura, S.; Itabashi, T.; Wada, Y.; Akimoto, M.; et al. Identifying pathogenic genetic background of simplex or multiplex retinitis pigmentosa patients: A large scale mutation screening study. *J. Med. Genet.* **2008**, *45*, 465–472. [CrossRef]

334. Jin, T.; Gu, Y.; Zanusso, G.; Sy, M.; Kumar, A.; Cohen, M.; Gambetti, P.; Singh, N. The Chaperone Protein bip binds to a mutant prion protein and mediates its degradation by the proteasome. *J. Biol. Chem.* **2000**, *275*, 38699–38704. [CrossRef] [PubMed]

335. Li, T.; Sandberg, M.A.; Pawlyk, B.S.; Rosner, B.; Hayes, K.C.; Dryja, T.P.; Berson, E.L. Effect of vitamin A supplementation on rhodopsin mutants threonine-17 –> methionine and proline-347 –> serine in transgenic mice and in cell cultures. *Proc. Natl. Acad. Sci. USA* **1998**, *95*, 11933–11938. [CrossRef] [PubMed]

336. Mattle, D.; Kuhn, B.; Aebi, J.; Bedoucha, M.; Kekilli, D.; Grozinger, N.; Alker, A.; Rudolph, M.G.; Schmid, G.; Schertler, G.F.X.; et al. Ligand channel in pharmacologically stabilized rhodopsin. *Proc. Natl. Acad. Sci. USA* **2018**, *115*, 3640–3645. [CrossRef] [PubMed]

337. Vasireddy, V.; Chavali, V.R.M.; Joseph, V.T.; Kadam, R.; Lin, J.H.; Jamison, J.A.; Kompella, U.B.; Reddy, G.B.; Ayyagari, R. Rescue of photoreceptor degeneration by curcumin in transgenic rats with P23H rhodopsin mutation. *PLoS ONE* **2011**, *6*, e21193. [CrossRef] [PubMed]

338. Chen, Y.; Chen, Y.; Jastrzebska, B.; Golczak, M.; Gulati, S.; Tang, H.; Seibel, W.; Li, X.; Jin, H.; Han, Y.; et al. A novel small molecule chaperone of rod opsin and its potential therapy for retinal degeneration. *Nat. Commun.* **2018**, *9*, 1976. [CrossRef]

339. Chiu, A.M.; Mandziuk, J.J.; Loganathan, S.K.; Alka, K.; Casey, J.R. High throughput assay identifies glafenine as a corrector for the folding defect in corneal dystrophy–causing mutants of SLC4A11. *Investig. Ophthalmol. Vis. Sci.* **2015**, *56*, 7739–7753. [CrossRef]

340. Robben, J.H.; Sze, M.; Knoers, N.V.A.M.; Deen, P.M.T. Rescue of vasopressin V2 receptor mutants by chemical chaperones: Specificity and mechanism. *Mol. Biol. Cell* **2006**, *17*, 379–386. [CrossRef]

MDPI

St. Alban-Anlage 66

4052 Basel

Switzerland

Tel. +41 61 683 77 34

Fax +41 61 302 89 18

www.mdpi.com

Biomolecules Editorial Office

E-mail: biomolecules@mdpi.com

www.mdpi.com/journal/biomolecules